WITHDRAWN

Selected Titles in This Series

727 **Anthony D. Blaom,** A geometric setting for Hamiltonian perturbation theory, 2001
726 **Victor L. Shapiro,** Singular quasilinearity and higher eigenvalues, 2001
725 **Jean-Pierre Rosay and Edgar Lee Stout,** Strong boundary values, analytic functionals, and nonlinear Paley-Wiener theory, 2001
724 **Lisa Carbone,** Non-uniform lattices on uniform trees, 2001
723 **Deborah M. King and John B. Strantzen,** Maximum entropy of cycles of even period, 2001
722 **Hernán Cendra, Jerrold E. Marsden, and Tudor S. Ratiu,** Lagrangian reduction by stages, 2001
721 **Ingrid C. Bauer,** Surfaces with $K^2 = 7$ and $p_g = 4$, 2001
720 **Palle E. T. Jorgensen,** Ruelle operators: Functions which are harmonic with respect to a transfer operator, 2001
719 **Steve Hofmann and John L. Lewis,** The Dirichlet problem for parabolic operators with singular drift terms, 2001
718 **Bernhard Lani-Wayda,** Wandering solutions of delay equations with sine-like feedback, 2001
717 **Ron Brown,** Frobenius groups and classical maximal orders, 2001
716 **John H. Palmieri,** Stable homotopy over the Steenrod algebra, 2001
715 **W. N. Everitt and L. Markus,** Multi-interval linear ordinary boundary value problems and complex symplectic algebra, 2001
714 **Earl Berkson, Jean Bourgain, and Aleksander Pełczynski,** Canonical Sobolev projections of weak type $(1, 1)$, 2001
713 **Dorina Mitrea, Marius Mitrea, and Michael Taylor,** Layer potentials, the Hodge Laplacian, and global boundary problems in nonsmooth Riemannian manifolds, 2001
712 **Raúl E. Curto and Woo Young Lee,** Joint hyponormality of Toeplitz pairs, 2001
711 **V. G. Kac, C. Martinez, and E. Zelmanov,** Graded simple Jordan superalgebras of growth one, 2001
710 **Brian Marcus and Selim Tuncel,** Resolving Markov chains onto Bernoulli shifts via positive polynomials, 2001
709 **B. V. Rajarama Bhat,** Cocylces of CCR flows, 2001
708 **William M. Kantor and Ákos Seress,** Black box classical groups, 2001
707 **Henning Krause,** The spectrum of a module category, 2001
706 **Jonathan Brundan, Richard Dipper, and Alexander Kleshchev,** Quantum Linear groups and representations of $GL_n(\mathbb{F}_q)$, 2001
705 **I. Moerdijk and J. J. C. Vermeulen,** Proper maps of toposes, 2000
704 **Jeff Hooper, Victor Snaith, and Min van Tran,** The second Chinburg conjecture for quaternion fields, 2000
703 **Erik Guentner, Nigel Higson, and Jody Trout,** Equivariant E-theory for C^*-algebras, 2000
702 **Ilijas Farah,** Analytic guotients: Theory of liftings for quotients over analytic ideals on the integers, 2000
701 **Paul Selick and Jie Wu,** On natural coalgebra decompositions of tensor algebras and loop suspensions, 2000
700 **Vicente Cortés,** A new construction of homogeneous quaternionic manifolds and related geometric structures, 2000
699 **Alexander Fel′shtyn,** Dynamical zeta functions, Nielsen theory and Reidemeister torsion, 2000
698 **Andrew R. Kustin,** Complexes associated to two vectors and a rectangular matrix, 2000
697 **Deguang Han and David R. Larson,** Frames, bases and group representations, 2000

(Continued in the back of this publication)

A Geometric Setting for Hamiltonian Perturbation Theory

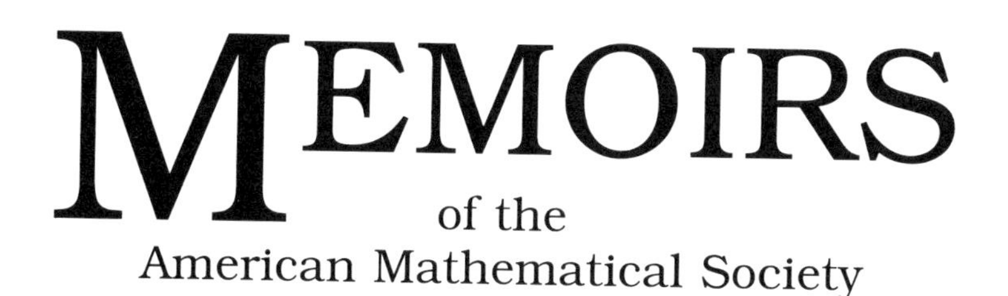

Number 727

A Geometric Setting for Hamiltonian Perturbation Theory

Anthony D. Blaom

September 2001 • Volume 153 • Number 727 (third of 5 numbers) • ISSN 0065-9266

American Mathematical Society
Providence, Rhode Island

2000 *Mathematics Subject Classification.*
Primary 70H09, 70H08, 37J15; Secondary 37J40, 70H33, 70H06.

Library of Congress Cataloging-in-Publication Data

Blaom, Anthony D., 1968–
A geometric setting for Hamiltonian perturbation theory / Anthony D. Blaom.
p. cm. — (Memoirs of the American Mathematical Society, ISSN 0065-9266 ; no. 727)
"September 2001, volume 153, number 727 (third of 5 numbers)."
Includes bibliographical references.
ISBN 0-8218-2720-0
1. Perturbation (Mathematics). 2. Hamiltonian systems. I. Title. II. Series.

QA3.A57 no. 727
[QA871]
510 s—dc21
[515′.35] 2001032797

Memoirs of the American Mathematical Society

This journal is devoted entirely to research in pure and applied mathematics.

Subscription information. The 2001 subscription begins with volume 149 and consists of six mailings, each containing one or more numbers. Subscription prices for 2001 are $494 list, $395 institutional member. A late charge of 10% of the subscription price will be imposed on orders received from nonmembers after January 1 of the subscription year. Subscribers outside the United States and India must pay a postage surcharge of $31; subscribers in India must pay a postage surcharge of $43. Expedited delivery to destinations in North America $35; elsewhere $130. Each number may be ordered separately; *please specify number* when ordering an individual number. For prices and titles of recently released numbers, see the New Publications sections of the *Notices of the American Mathematical Society.*

Back number information. For back issues see the *AMS Catalog of Publications.*

Subscriptions and orders should be addressed to the American Mathematical Society, P. O. Box 845904, Boston, MA 02284-5904. *All orders must be accompanied by payment.* Other correspondence should be addressed to Box 6248, Providence, RI 02940-6248.

Copying and reprinting. Individual readers of this publication, and nonprofit libraries acting for them, are permitted to make fair use of the material, such as to copy a chapter for use in teaching or research. Permission is granted to quote brief passages from this publication in reviews, provided the customary acknowledgment of the source is given.

Republication, systematic copying, or multiple reproduction of any material in this publication is permitted only under license from the American Mathematical Society. Requests for such permission should be addressed to the Assistant to the Publisher, American Mathematical Society, P. O. Box 6248, Providence, Rhode Island 02940-6248. Requests can also be made by e-mail to reprint-permission@ams.org.

Memoirs of the American Mathematical Society is published bimonthly (each volume consisting usually of more than one number) by the American Mathematical Society at 201 Charles Street, Providence, RI 02904-2294. Periodicals postage paid at Providence, RI. Postmaster: Send address changes to Memoirs, American Mathematical Society, P. O. Box 6248, Providence, RI 02940-6248.

© 2001 by the American Mathematical Society. All rights reserved.

This publication is indexed in *Science Citation Index®, SciSearch®, Research Alert®, CompuMath Citation Index®, Current Contents®/Physical, Chemical & Earth Sciences.*
Printed in the United States of America.

∞ The paper used in this book is acid-free and falls within the guidelines established to ensure permanence and durability.
Visit the AMS home page at URL: http://www.ams.org/

10 9 8 7 6 5 4 3 2 1 06 05 04 03 02 01

Contents

Abstract

The perturbation theory of non-commutatively integrable systems is revisited from the point of view of non-Abelian symmetry groups. Using a coordinate system intrinsic to the geometry of the symmetry, we generalize and geometrize well-known estimates of Nekhoroshev (1977), in a class of systems having almost G-invariant Hamiltonians. These estimates are shown to have a natural interpretation in terms of momentum maps and co-adjoint orbits. The geometric framework adopted is described explicitly in examples, including the Euler-Poinsot rigid body.

Received by the editor May 26, 1998 and in revised form April 14, 2000.

The author thanks Jerry Marsden for considerable help and support during this project. Helpful feedback has been supplied by Francesco Fassò, Tudor Ratiu and Pierre Lochak.

Key words and phrases. Hamiltonian perturbation theory, non-Abelian symmetry, action-angle coordinates, non-canonical coordinates, rigid body, Nekhoroshev estimate, momentum map, moment map, co-adjoint orbit, symplectic cross-section, action-group coordinates, Hamiltonian G-space.

2000 Mathematics Subject Classification. Primary 70H09 70H08 37J15 ; Secondary 37J40 70H33 70H06.

Notation

$\mathbb{T}^n$	n-dimensional torus
G	compact connected Lie group;
	in Chap. 5 and Appendix A, a compact real-analytic manifold;
	in Chap. 6, a compact connected *semisimple* Lie group;
n_G	order of the matrix group $\mathbb{R}^{n_G \times n_G}$ in which G is embedded (p. 10)
$G^{\mathbb{C}}$, $(G^\sigma)_{0 \leqslant \sigma \leqslant \bar\sigma}$	complexification of G (p. 10), local complexification of G (p. 27,32)
$\mathfrak{g}$	Lie algebra of G
$\mathfrak{g}_{\mathrm{reg}}$, $\mathfrak{g}^*_{\mathrm{reg}}$	set of regular points of adjoint (resp. co-adjoint) action (p. 8)
$T \subset G$	fixed maximal torus
$\mathfrak{t}$	Lie algebra of T;
	in Chap. 5 and Appendix A, an arbitrary real vector space
$\mathfrak{t}^{\mathbb{Z}}$	period lattice of Assumption A (p. 28);
	in Chap. 12, the integral lattice of T in $\mathfrak{t}$
k	dimension of $\mathfrak{t}$
$V^{\mathbb{C}}$, $\operatorname{Ann} V$	complexification and annihilator of a space V
$a \cdot b$	fixed Ad-invariant inner product on $\mathfrak{g}$, extended
	to a $\mathbb{C}$-bilinear form on $\mathfrak{g}^{\mathbb{C}}$ (p. 9);
	in Chap. 6, minus the Killing form (p. 31);
	in Chap. 5 and Appendix A, an arbitrary inner product on $\mathfrak{t}$,
	and its extension to $\mathfrak{t}^{\mathbb{C}}$ (p. 27)
$\mathfrak{t}^\perp = [\mathfrak{g}, \mathfrak{t}]$	orthogonal of $\mathfrak{t}$ in $\mathfrak{g}$ (pp. 8,9)
$\underline{\mathfrak{t}}$, $\underline{\mathfrak{t}}^\perp$	annihilators of $\mathfrak{t}^\perp$, $\mathfrak{t}$ (p. 9)
φ	isomorphism $\mathfrak{g} \to \mathfrak{g}^*$ induced by Ad-invariant
	inner product on $\mathfrak{g}$ (p. 9)
i	restriction $\underline{\mathfrak{t}} \to \mathfrak{t}^*$ of natural projection $\mathfrak{g}^* \to \mathfrak{t}^*$ (p. 10)
$\mathfrak{t}_0$	Weyl chamber (p. 8);
	in Chap. 5 and Appendix A, a certain open subset of $\mathfrak{t}$ (p. 27)
$\mathcal{W}$	Weyl chamber in $\mathfrak{g}^*$ (p. 9)
$\mathfrak{t}_0^* \equiv i(\mathcal{W})$	Weyl chamber (p. 10)
ω_G^*, ω_G	symplectic structure on $G \times \mathfrak{t}_0^*$ (resp. $G \times \mathfrak{t}_0$) (pp. 10,11,54)
Θ_G^*, Θ_G	one-form on $G \times \mathfrak{t}_0^*$ (resp. $G \times \mathfrak{t}_0$) with
	$d\Theta_G^* = \omega_G^*$ (resp. $d\Theta_G = \omega_G$) (pp. 11,54)
$(\xi_0, \tau_0)_{(g,p)}$	see p. 11
$\xi \cdot \frac{\partial}{\partial g}$, $\tau \cdot \frac{\partial}{\partial p}$	see p. 11
$\mathfrak{t}_0^{\mathbb{C}}$	see p. 11
λ_p	inverse of $\mathrm{ad}_p : \mathfrak{t}^{\perp\mathbb{C}} \to \mathfrak{t}^{\perp\mathbb{C}}$ (p. 11);
	in Part 2, inverse of $\xi \mapsto \mathrm{ad}^*_\xi(i^{-1}(p)) : \mathfrak{t}^\perp \to \underline{\mathfrak{t}}^\perp$ (p. 55)
ξ_H, τ_H	see p. 12
$\frac{\partial H}{\partial g}$, $\frac{\partial H}{\partial p}$	see pp. 12,56

σ	projection $\mathfrak{g}^{\mathbb{C}} \to \mathfrak{t}^{\mathbb{C}}$ along $\mathfrak{t}^{\perp\mathbb{C}}$ (p. 12);
$\sigma^{\perp}$	projection $\mathfrak{g}^{\mathbb{C}} \to \mathfrak{t}^{\perp\mathbb{C}}$ along $\mathfrak{t}^{\mathbb{C}}$ (p. 12)
σ^*	projection $\mathfrak{g}^* \to \underline{\mathfrak{t}}$ along $\underline{\mathfrak{t}}^{\perp}$ (p. 56)
$\hat{\xi}$	element of $\mathfrak{so}(3, \mathbb{C})$ defined through $\hat{\xi} u = \xi \times u$, for $\xi, u \in \mathbb{C}^3$ (p. 13)
$\{u, v\}$	Poisson bracket of u and v (p. 13)
ξ_M	infinitesimal generator of action of G on M, along some $\xi \in \mathfrak{g}$
P	symplectic manifold
$\mathbf{J}$	momentum map of G-action on P
J_ξ	ξ-component of $\mathbf{J}$ (p. 15)
$\mathbf{J}^G$	momentum map of G-action on $G \times \mathfrak{t}_0$ (p. 16); in Part 2, momentum map of G-action on $G \times \mathfrak{t}_0^*$ (p. 55)
$\mathbf{j}^G$	momentum map of T-action on $G \times \mathfrak{t}_0$ (p. 22); in Part 2, momentum map of T-action on $G \times \mathfrak{t}_0^*$ (p. 55)
Ω	frequency $(=\nabla h)$ (p. 21)
I_T	integral lattice of torus T in $\mathfrak{t}$ (p. 22)
$\bar{H}$	T-average of H (p. 23)
$\operatorname{int} M$	interior of M
$\mathcal{A}^V(M)$	space of continuous maps $u : M \to V$ that are real-analytic (resp. holomorphic) on $\operatorname{int} M$, for a real (resp. complex) vector space V (p. 26)
$\\|\cdot\\|$	supremum norm on $\mathcal{A}^V(M)$ (p. 26)
$B_R(p)$	closed R-ball in $\mathfrak{t}$ with center p (p. 27)
$B_R^{\rho}(p)$	closed ρ-neighborhood of $B_R(p)$ in $\mathfrak{t}^{\mathbb{C}}$($p$. 27)
$\bar{\sigma},\ \bar{\rho}$	Hamiltonian analyticity widths (p. 27)
$\bar{p}$	fixed element of $\mathfrak{t}_0$ (or $U \subset \mathfrak{t}_0$ (pp. 27,42)
$p_*,\ r,\ \gamma$	variable parameters ($p_* \in \operatorname{int} B$) (p. 28)
$B,\ B^{\bar{\rho}}$	shorthand for $B_{\bar{\rho}}(\bar{p})$ and $B_{\bar{\rho}}^{\bar{\rho}}(\bar{p})$ (see above)
$D_\gamma(p_*, \rho)$	the domain $G^{\gamma\bar{\sigma}} \times B_{r\bar{\rho}}^{\gamma r\bar{\rho}}(p_*)$ (p. 28)
$\\|\cdot\\|_\gamma^{p_*,r}$	supremum norm on $\mathcal{A}^{\mathbb{C}}(D_\gamma(p_*, r))$ (p. 28)
$T_m, T_M, T_\Omega, T_{h'''}$	time constants (p. 29)
$c_1, \ldots, c_7$	dimensionless constants (pp. 28–30)
E	energy constant $(= \bar{\sigma}\bar{\rho}/T_M)$ (p. 30)
$\|\cdot\|_S$	norm on $\mathbb{C}^{n_G \times n_G}$ induced by Schur-Hadamard product S (p. 32)
k_1, k_2, k_3	dimensionless constants (pp. 32,36)
$\bar{k}$	number of positive roots of $\mathfrak{g}$ in $\mathfrak{t}^*$ (p. 32)
$\alpha_1, \ldots \alpha_{\bar{k}}$	positive real roots of $\mathfrak{g}$ in $\mathfrak{t}^*$ (p. 32)
κ	isomorphism $\mathfrak{t} \to \mathfrak{t}^*$ induced by Killing form (p. 33)
$B(\cdot, \cdot)$	inner product on $\mathfrak{t}^*$ induced by Killing form (p. 33)
$\alpha_1^*, \ldots, \alpha_{\bar{k}}^*$	inverse roots $\mathfrak{g}$ in $\mathfrak{t}$ (p. 33)
n_{ij}	Cartan integers of $\mathfrak{g}$ (p. 33)
m_{ij}	see p. 33
ζ	dimensionless constant (p. 39)
$M \backslash Z$	set of points in M not in Z
$\pi_{\mathcal{O}}$	projection $\mathfrak{g}^*_{\mathrm{reg}} \to \mathcal{O}$ ($\mathcal{O}$ a co-adjoint orbit; p. 47)
$\pi_{\mathcal{W}}$	projection $\mathfrak{g}^*_{\mathrm{reg}} \to \mathcal{W}$ (p. 47)

π	symplectic fibration $P \to \mathcal{O}$ (pp. 47,48)
F_ν	fiber $\pi^{-1}(\nu)$ (p. 48)
F_μ	pseudoleaf $\mathbf{j}^{-1}(\mu)$ (p. 70)
F	symplectic cross-section $\pi^{-1}(\mu_0)$ of P (p. 48), or arbitrary Hamiltonian T-space
ω_F	symplectic structure on F (p. 48)
$\mathbf{J}^F$	momentum map $F \to \mathfrak{t}^*$ of action of T on F (p. 48)
hor_x	horizontal space at $x \in P$ of symplectic connection on $\pi : P \to \mathcal{O}$ (p. 49)
Θ	canonical one-form on T^*G (p. 53)
$L_g,\ R_g$	map $G \to G$ sending g to gh (resp. hg)
$\mathrm{T}_*\phi$	covariant cotangent lift of ϕ (p. 53)
$(\xi, \tau)_{\mathrm{vf}}$	vector field on $G \times \mathfrak{t}_0^*$ defined on p. 54
Λ_p	map $\mathfrak{g}^* \to \mathfrak{t}^\perp$ defined on p. 56
$\boldsymbol{\lambda},\ \boldsymbol{\rho}$	left (resp. right) trivialization $\mathrm{SO}(3) \times \mathbb{R}^3 \xrightarrow{\sim} \mathrm{T}\,\mathrm{SO}(3)$ (p. 65)
ρ	projection $F \to F/T$ (p. 70)
ω_μ	symplectic form on leaf F_μ (p. 70)
D	arbitrary complement for the characteristic distribution on F/T (p. 71)
L_y	natural isomorphism $\mathfrak{t}^* \to D(y)$ (p. 71)
D_μ	differential operator $\Omega^k(F/T;\mathbb{R}) \to \Omega^k(F_\mu;\mathfrak{t})$ defined on p. 71

Overture

A concise exposition of the results to be expounded in this memoir will be offered in the Introduction. While some of these results are sketched here, the main task at hand is to describe, informally and in plain language, the questions and problems motivating our work, as well as the ideas and point of view leading us ultimately to answers. It will be illustrative to keep two examples in mind that serve as good prototypes for the larger class of systems that will be studied later.

Two simple Hamiltonian systems

Example A. Consider the motion of a point mass m moving through ordinary three-dimensional space, free of external forces including gravity. The linear momentum $\mathbf{J}_{\text{lin}}$ of the mass is conserved, so that the motion is in a straight line at constant speed.

Example B. Alternatively, suppose that our point mass is constrained to move on the surface of a smooth sphere, the only external force being the normal force necessary to maintain the constraint. In this case the angular momentum $\mathbf{J}_{\text{ang}}$ is conserved and the motion is along great circles of the sphere at constant speed.

Both Examples A and B possess a *conservation law*. The conservation laws can be 'explained' by the existence of *symmetries* in the underlying equations. This is the content of a well-known theorem of Noether (see, e.g., Marsden (1992)). Example A clearly possesses a three-dimensional *translational* symmetry, while Example B possesses a *rotational* symmetry.

Suppose these symmetries are broken. For instance, suppose that in Example A we add a weak spatially periodic potential, and in Example B we introduce a small aspherical distortion. What is the fate of the above conservation laws?

The complexity of perturbed motions

In the absence of perturbations, Examples A and B are both *integrable* systems, i.e., systems whose solutions can be written down in an explicit or 'closed' form. However, integrability is the exception rather than the rule, and generic perturbations to integrable systems create extraordinarily complicated behavior. See, e.g., Lichtenberg and Lieberman (1992). In particular, Poincaré's nonexistence theorem (see, e.g., Benettin, Ferrari, Galgani and Giorgilli (1982)) tells us that conserved quantities in an integrable system are irrevocably destroyed by generic perturbations. Despite this, however, there are two important results establishing, under appropriate hypotheses, some kind of 'stability' for the conserved quantities. These are the Kolmogorov-Arnold-Moser (or KAM) theorem (Kolmogorov, 1954; Arnold, 1963; Moser, 1962) and Nekhoroshev's theorem (Nekhoroshev, 1977). For generic perturbations, the KAM theorem does not directly apply to Example B (or to other examples like it), and we discuss it no further[1]. Nekhoroshev's theorem can be applied to both Examples A and B, but not without significant differences that we now elucidate.

Nekhoroshev's theorem

Nekhoroshev's theorem applies to systems admitting *action-angle coordinates.* Nekhoroshev also stated his result in terms of a generalization called *partial action-angle coordinates*, but his argument in the more general setting was incomplete (see below). These coordinates are relevant in Example B, but not in Example A, which we describe first.

Action-angle coordinates in Example A are easily constructed. Suppose as before, that attention is restricted to spatially periodic perturbations, and let L be a 'lattice of periodicity.' For simplicity, assume that L has an orthogonal basis, and that the corresponding periods of L are the same in each direction. Nondimensionalize all lengths by this period, and all masses by the mass of m. Modulo the lattice L, the position of m is determined by three numbers q_1, q_2, q_3 with $0 \leqslant q_j < 1$; these numbers constitute the 'angles.'[2] The associated 'actions' are the components p_1, p_2, p_3 of the linear momentum $\mathbf{J}_{\text{lin}}$. What makes these coordinates action-angle coordinates is that: (i) they put the equations of motion into *Hamiltonian form,*

$$\dot{q}_j = \frac{\partial H}{\partial p_j} , \qquad \dot{p}_j = -\frac{\partial H}{\partial q_j} ,$$

where the appropriate *Hamiltonian* H is in this case the total energy of the system, and (ii) the unperturbed value H_0 of the Hamiltonian (i.e., the kinetic energy) depends only on the action variables:

$$H_0(q,p) = h(p) \equiv \frac{1}{2}p_1^2 + \frac{1}{2}p_2^2 + \frac{1}{2}p_3^2 .$$

In particular, the unperturbed motions of a system in action-angle coordinates are quasiperiodic:

$$p_j(t) = p_j(0) , \qquad q_j(t) = q_j(0) + t\Omega_j ,$$

[1] The interested reader is referred to, e.g., Arnold, Kozlov and Neishtadt (1988).

[2] The familiar scaling $0 \leqslant q_j < 2\pi$ produces distracting factors of 2π later on.

where $\Omega_j \equiv \partial h/\partial p_j(p(0))$. In our example, we have $\Omega_j = p_j(0)$, and the constancy of the action variables corresponds to the conservation of momentum $\mathbf{J}_{\text{lin}}$.

We are now ready to apply Nekhoroshev's result:

Nekhoroshev's Theorem. (Nekhoroshev (1977)[3]) *Consider a Hamiltonian system in action-angle coordinates* $q_1, \dots, q_n$, $p_1, \dots, p_n$ *and consider perturbed Hamiltonians of the form*

$$H(q,p) = h(p) + \epsilon F(q,p) \ .$$

Restrict attention to values of the action vector $p \equiv (p_1, \dots, p_n)$ *lying in some finite closed ball* B, *and initial conditions* $p(0)$ *lying in the interior of* B. *Assume* h *and* F *are real-analytic, and that the level sets in* B *of the unperturbed Hamiltonian* h *are strictly convex. Then there exist positive constants* a, b, c, t_0 *and* r_0 *such that for all sufficiently small* $\epsilon \geqslant 0$ *all solutions of the perturbed system satisfy the exponential estimate*

$$|t| \leqslant t_0 \exp(c\epsilon^{-a}) \quad \Longrightarrow \quad |p(t) - p(0)| \leqslant r_0 \epsilon^b \ .$$

In Example A we have $p = \mathbf{J}_{\text{lin}}$, so that order ϵ perturbations lead to order ϵ^b 'drifts' in the momentum $\mathbf{J}_{\text{lin}}$ of the mass m after times on the order of $\exp(c\epsilon^{-a})$, at most. If ϵ is small, such times can be very large indeed. For example, in an application by Giorgilli and Skokos (1997) to Trojan asteroids, the Nekhoroshev-type stability times derived are on the order of the current estimated age of the universe!

Non-commutative integrability

Before turning to Example B, let us summarize some key observations about Example A, which are typical of systems admitting action-angle coordinates:

A1. Unperturbed motions in Example A (which has *three* degrees of freedom) are quasiperiodic with *three* independently controllable[4] frequencies.

A2. The conserved quantity $\mathbf{J}_{\text{lin}}$ is a vector with *three* components.

A3. The underlying translational symmetry is *Abelian*, meaning that the net effect of two successive translations is independent of the order in which they are applied.

We add one final observation which is less obvious but nevertheless important:

A4. The Poisson bracket (see below) $\{J_i, J_j\}$ of any pair of components J_1, J_2, J_3 of $\mathbf{J}_{\text{lin}}$ vanishes.

By definition, the *Poisson bracket* $\{f, h\}$ of two functions f and h is computed by differentiating f along solution curves of the system obtained by taking as Hamiltonian the function h.

If we are to apply Nekhoroshev's theorem as above to Example B, then we shall first need to construct action-angle coordinates. But at odds with this requirement is the disturbing fact that analogues of the observations A1–A4 follow an altogether different pattern in Example B:

[3] For this formulation see, e.g., Lochak (1992). In Nekhoroshev's original statement, the convexity condition is replaced by a 'steepness' condition. This is a weaker (indeed C^∞-generic) condition we will not attempt to consider here or elsewhere.

[4] Controllable by varying the initial conditions $p_j(0)$.

B1. Unperturbed motions in Example B (which has *two* degrees of freedom) are periodic, i.e., have only *one* associated frequency.
B2. The conserved quantity $\mathbf{J}_{\text{ang}}$ has *three* components, i.e., one more component than the system has degrees of freedom.
B3. The underlying rotational symmetry is *non-Abelian*, since the net effect of applying successive rotations in three-space depends in general on the order in which these rotations are applied.
B4. The Poisson brackets of components J_1, J_2, J_2 of $\mathbf{J}_{\text{ang}}$ satisfy the cyclic conditions

$$\{J_1, J_2\} = J_3 \ , \qquad \{J_2, J_3\} = J_1 \ , \qquad \{J_3, J_1\} = J_2 \ .$$

What is remarkable is that (local) action-angle coordinates *can* be constructed in Example B although, as one might imagine in view of the incongruence of the properties listed above, the construction is substantially more complicated than was the case in Example A. We do not attempt to describe this construction here. At any rate, the unperturbed Hamiltonian $H_0(q,p) = h(p)$ in action-angle coordinates fails to be convex, so that Nekhoroshev's theorem above fails to apply.

Integrable systems whose conserved quantities satisfy nontrivial Poisson bracket relations, as in B4, are known as *non-commutatively integrable* systems. In work preceeding the research on exponential estimates, Nekhoroshev (1972) determined that a large class of non-commutatively integrable systems admit a generalization of action-angle coordinates known as *partial action-angle coordinates*, which are in some sense more natural. Partial action-angle coordinates consist of two sets of variables: The first set consists of k angles $q_1, \ldots, q_k$ and k conjugate actions $p_1, \ldots, p_k$. The second set consists of $(n-k)$ variables $x_1, \ldots, x_{n-k}$ and $(n-k)$ conjugate variables $y_1, \ldots, y_{n-k}$, which are all ordinary (i.e., non-angular) coordinate functions. Here n denotes the total number of degrees of freedom. These coordinates, like conventional action-angle coordinates, are *canonical* in the sense that they put the equations of motion into Hamiltonian form:

$$\dot{q}_j = \frac{\partial H}{\partial p_j} \ , \quad \dot{p}_j = -\frac{\partial H}{\partial q_j} \ ; \qquad \dot{x}_j = \frac{\partial H}{\partial y_j} \ , \quad \dot{y}_j = -\frac{\partial H}{\partial x_j} \ .$$

In partial action-angle coordinates, a non-commutatively integrable Hamiltonian depends only on the p_j variables.

In Nekhoroshev's work on exponential estimates, he realized that most of his arguments carry over to the non-commutatively integrable case if one uses partial action-angle coordinates, and assumes convexity of the unperturbed Hamiltonian with respect to the p_j variables only. Even in Example B, however (the simplest example of a non-commutatively integrable system imaginable!), constructing partial action-angle coordinates is not trivial. Furthermore, one cannot construct partial action-angle coordinates globally in Example B without coordinate singularities. This is true even if we remove points in phase space corresponding to trivial motions (the mass m at rest), which constitute a 'natural' singularity of the problem.

Another problem with Nekhoroshev's generalized setting, as Fassò (1995) has pointed out, is that 'fast' (i.e., order ϵ) motions in the x_j, y_j variables take solutions to the perturbed problem in a non-commutatively integrable system out of locally defined partial action-angle coordinate charts, before the exponential estimates on the p_j variables can be rigorously established. Fassò overcomes this problem by showing how to make intrinsic sense of 'normal forms' for the perturbed Hamiltonian, although to express and compute Nekhoroshev type estimates still requires one to fix an atlas of partial action-angle coordinate charts. The extent to which these estimates depend on the choice of atlas is an open problem (Fassò, 1995, Appendix C).

The symmetry point of view

The reason for the difficulties presented by action-angle or partial action-angle coordinates in non-commutatively integrable systems is that the canonical nature of such coordinates is at odds with the intrinsic non-Abelian symmetry underlying many such systems. *We shall take the point of view that the geometry of an underlying symmetry, as well as its corresponding conservation law, should be built into whatever geometric framework is employed to carry out an analysis of perturbations.* This viewpoint is to take precedence over previous requirements that one work exclusively with *canonical* coordinate systems. Rather, one should endeavor to understand how non-canonical contributions enter the equations of motion, when these are viewed in a coordinate system intrinsic to the non-Abelian symmetry.

With this kind of understanding in hand, we seek to *geometrize* and *generalize* the commutative version of Nekhoroshev's theorem given above, in such a way that its application to systems with a perturbed non-Abelian symmetry becomes transparent. In particular, this generalization should yield estimates that apply immediately to the conservation law intrinsically associated with such a symmetry.

Action-group coordinates

The simplest coordinate system of the kind we are advocating is one that we shall refer to as action-*group* coordinates. In Example A, and other systems whose integrability arises from the existence of an appropriate Abelian symmetry, action-group coordinates correspond to conventional action-angle coordinates. A discussion of action-group coordinates in general is postponed to Chap. 2. We preview these coordinates here in the special case of Example B, which we shall revisit throughout our exposition of the general theory. A more sophisticated demonstration of this theory (Chap. 11) will be an application to the Euler-Poinsot rigid body.

Nondimensionalize all lengths in Example B by the radius of the sphere, and all masses by that of m. The state of the system is determined by the position q and instantaneous linear momentum vector v of m. We view q and v and vectors in three-space, subject to the conditions $\|q\| = 1$ and $v \cdot q = 0$ consistent with the physical constraints.

Assume the unit sphere is centered on the origin of an inertial orthonormal frame with basis e_1, e_2, e_3. It is convenient to express states (q, v) of m with respect to a *reference state* (q_0, v_0), which we choose arbitrarily to be (e_1, e_2). For an arbitrary state (q, v) there exists a 3×3 orientation preserving rotation matrix g and a number $p \geqslant 0$ such that

$$q = gq_0 \ , \qquad v = pgv_0 \ .$$

If we ignore trivial states ($v = 0$), then g and p are determined *uniquely*, and p is strictly positive. Whence g, p constitute 'coordinates,' in the sense that there is a one-to-one correspondence between states (q, v) and values for the pair (g, p).

Notice that the rotational symmetry of the unperturbed system is completely transparent in the coordinates (g, p) as the unperturbed Hamiltonian $H_0(g, p)$ (i.e., the kinetic energy determined by the standard metric on the sphere[5]) is independent of the 'symmetry' coordinate g; it depends only on the 'action' coordinate p, in the spirit of constructions of conventional action-angle coordinates:

$$H_0(g, p) = h(p) \equiv \frac{1}{2}p^2 \ .$$

The equations of motion take the form

$$(*) \qquad \dot{g} = g \begin{bmatrix} \frac{1}{p}\frac{\partial H}{\partial g_2} \\ -\frac{1}{p}\frac{\partial H}{\partial g_1} \\ \frac{\partial H}{\partial p} \end{bmatrix}^{\wedge} , \qquad \dot{p} = -\frac{\partial H}{\partial g_3} ,$$

where

$$\begin{bmatrix} \xi_1 \\ \xi_2 \\ \xi_3 \end{bmatrix}^{\wedge} \equiv \begin{bmatrix} 0 & -\xi_3 & \xi_2 \\ \xi_3 & 0 & -\xi_1 \\ -\xi_2 & \xi_1 & 0 \end{bmatrix}$$

and

$$\frac{\partial H}{\partial g_j}(g, p) \equiv \frac{d}{dt} H(ge^{t\hat{e}_j}, p)\Big|_{t=0} \qquad (1 \leqslant j \leqslant 3)$$

$$\frac{\partial H}{\partial p}(g, p) \equiv \frac{d}{dt} H(g, p + t)\Big|_{t=0} \ .$$

Roughly speaking, the non-canonical contributions to the equations are the terms

$$\frac{1}{p}\frac{\partial H}{\partial g_2} , \qquad -\frac{1}{p}\frac{\partial H}{\partial g_1} \ .$$

In the unperturbed case, the equations reduce to

$$\dot{g} = g(pe_3)^{\wedge} , \qquad \dot{p} = 0 ,$$

which admit the general solution

$$p(t) = p(0) \ , \qquad g(t) = g(0)\exp(t\Omega\hat{e}_3) \ ,$$

where $\Omega \equiv p(0)$.

Nekhoroshev estimates in the non-commutative case

The angular momentum of m in Example B is given by

$$\mathbf{J}_{\mathrm{ang}} = q \times v = pg(q_0 \times v_0) = pge_3 \ .$$

If one keeps $p > 0$ constant and runs through all possible values of the symmetry variable g, then the corresponding locus of $\mathbf{J}_{\mathrm{ang}}$ in three-space is a sphere of radius p. In fact these spheres are intrinsic geometric objects associated with the rotational symmetry of the problem known as *co-adjoint orbits*. These orbits (which are zero dimensional in the Abelian case) influence the perturbed motions, and in particular the fate of the conservation law $\dot{\mathbf{J}}_{\mathrm{ang}} = 0$, in a fundamental way. As we show, Nekhoroshev type estimates for a generic perturbation apply to 'drifts' in the momentum $\mathbf{J}_{\mathrm{ang}}$ in those directions *transverse* to the co-adjoint orbits (spheres). That is, Nekhoroshev estimates, under appropriate hypotheses, apply to the action

[5]We realize perturbations in the form of aspherical distortions by perturbing this metric, not by perturbing the sphere with which configurations q are identified.

variable $p = \|\mathbf{J}_{\text{ang}}\|$. Relatively fast motions (order ϵ) can and do appear in the *tangential* directions, being projections onto three-space of fast dynamics in the g variable. These motions correspond to those pointed out by Fassò and discussed above. Unlike partial action-angle coordinate charts however, an action-group coordinate chart completely contains such motions.

Introduction

Hamiltonian perturbation theory begins with the construction of a symplectic diffeomorphism between some open subset of the phase space P and an appropriate open subset of a 'model space' M. Presently, action-angle coordinates ($M = \mathbb{T}^n \times \mathbb{R}^n$) and partial action-angle coordinates ($M = \mathbb{T}^k \times \mathbb{R}^k \times \mathbb{R}^{2(n-k)}$, $0 \leqslant k < n$) are the exclusive choices. Both model spaces are *canonical* in the sense that, up to a covering, they are Euclidean space equipped with its usual symplectic structure. Which space is appropriate depends on the nature of the integrability of the unperturbed system.

Integrability

Classical notions of integrability, cast in terms of integrals of motion and their Poisson brackets (see, e.g., Arnold et al. (1988)), can be reformulated in geometric language (Dazord and Delzant, 1987): A Hamiltonian H_0, defined on some symplectic manifold (P, ω), is *integrable* if H_0 is constant on the leaves of a coisotropic and symplectically complete foliation $\mathcal{F}$ on P. A foliation $\mathcal{F}$ is *coisotropic* if its tangent distribution D contains its symplectic orthogonal D^ω. It is *symplectically complete* if D^ω is integrable as a distribution. In that case, the leaves of the corresponding foliation $\mathcal{F}^\omega$ are invariant under the flow of the Hamiltonian vector field X_{H_0}.

To the above 'differential' notion of integrability one adds a global condition.[6] The simplest of these, compactness of the leaves of $\mathcal{F}^\omega$, implies these leaves are actually k-dimensional tori and that $\mathcal{F}^\omega$ is an 'angular fibering' in the sense of Nekhoroshev (1972). In particular, in the case of *commutative* integrability ($k = n \equiv (\dim P)/2$) a neighborhood of each k-torus admits action-angle coordinates, while in the case of *non-commutative* integrability ($k < n$) a neighborhood of each such torus admits partial action-angle coordinates. In either case, the Hamiltonian H_0 is represented by a function depending only on the k actions 'conjugate' to the angles representing the tori. This has the consequence that X_{H_0}, restricted to a given k-torus is a *linear* vector field (i.e., is 'covered' by a linear vector field on $\mathbb{R}^k$). Proofs of these assertions and further details are given in Dazord and Delzant (1987).

Generally, in the non-commutative case, it is not possible to construct a partial action-angle coordinate chart such that it contains a full neighborhood of a leaf of $\mathcal{F}$. Unfortunately, as Fassò (1995) has pointed out (see also Benettin and Fassò (1996)), perturbations to H_0 create 'fast' motions (motions of the same order as the perturbation) along the leaves of $\mathcal{F}$. These motions take trajectories out of a locally defined partial action-angle coordinate chart in a relatively short time. Fassò

[6] Locally, *all* systems are integrable in the stated sense, away from equilibria: Simply apply Darboux's theorem, taking as first coordinate the Hamiltonian function.

overcomes this problem by showing how to make intrinsic sense of normal forms for the perturbed Hamiltonian. Such normal forms are used to deduce Nekhoroshev estimates on other components of the motion, although to compute and express such estimates a particular atlas of partial action-angle coordinate charts must be fixed. The extent to which these estimates depend on the choice is an open problem (see Fassò (1995, Appendix C)).

Another approach to non-commutative integrability, due to Mishchenko and Fomenko (1978) (see also Arnold et al. (1988)), is to convert non-commutative integrability into *commutative* integrability. This is accomplished, at least locally, by 'gluing' together the k-tori to form n-tori; out of these n-tori one builds conventional action-angle coordinates. There is no escaping the difficulties mentioned above, however, because the unperturbed Hamiltonian will still depend on only $k < n$ actions, leading to 'fast' motions in the remaining variables which take trajectories out of locally defined coordinate charts as before. Moreover, this degeneracy of the Hamiltonian appears mysteriously because its source, namely the non-commutative geometry of the original problem, is concealed by the 'commutative' coordinates employed.

Integrability through symmetry

Non-commutative integrability arises frequently in applications and is often manifest in the form of a non-Abelian symmetry group G. In that case the leaves of $\mathcal{F}$ correspond to orbits of G in the phase space[7]. Generally the geometry of the symmetry is obscured in partial action-angle coordinate charts since, as we remarked above, such charts need not contain full neighborhoods of leaves of $\mathcal{F}$[8]. An alternative that we explore here is to substitute for a canonical model space a *Hamiltonian G-space normal form*. One can view these normal forms as 'non-canonical coordinates' built directly out of the non-Abelian group action. The simplest of these normal forms is a generalization of action-angle coordinates that we shall refer to as *action-group coordinates* ($M = G \times \mathfrak{t}_0^*$; see Chap. 1). The 'group' in action-group coordinates can be any compact connected Lie group G; when $G = \mathbb{T}^n$, one recovers conventional action-angle coordinates. In action-group coordinates the k-tori discussed above are represented by cosets in G of some maximal torus $T \subset G$.

Action-group coordinates appear in Dazord and Delzant (1987, Section 5). Forerunners of these coordinates have been obtained by Marsden (1981), Gotay (1982), Marle (1983a) and Guillemin and Sternberg (1984, §41). We shall see that in the perturbation analysis of a non-Abelian symmetry, a single action-group coordinate chart suffices, and applies under conditions that are readily verified.

Generalized Nekhoroshev estimates

Our main objective is to demonstrate that a Hamiltonian G-space normal form can indeed be used as a geometric framework for perturbation theory. We do this in Part 1 by generalizing a Nekhoroshev's (1977) theorem to a 'non-canonical' setting that includes action-group coordinates (Theorem 5.8, p. 30). As a corollary

[7] A proof of this fact will be recalled in Remark 3.12.

[8] In fact (see Remark 4.4), if G acts freely and is compact connected and non-Abelian, then it is *never* possible for a partial action-angle coordinate chart to contain a neighborhood of a G-orbit (i.e., leaf of $\mathcal{F}$).

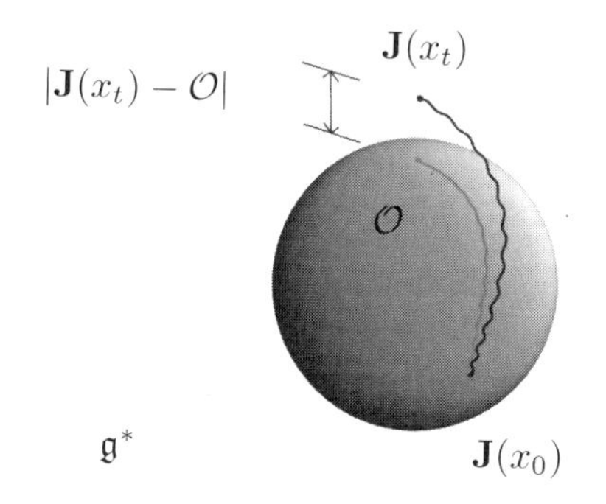

FIGURE 1. Schematic representation of a solution curve $t \mapsto x_t$ of the perturbed Hamiltonian H, projected onto 'momentum space' $\mathfrak{g}^*$ using the momentum map $\mathbf{J} : P \to \mathfrak{g}^*$. The co-adjoint orbit through the initial point $\mathbf{J}(x_0)$ (denoted $\mathcal{O}$) is depicted as a sphere.

we deduce a Nekhoroshev-type estimate on the evolution of momentum maps, in a class of integrable Hamiltonian systems with nearly G-invariant Hamiltonians (Corollary 7.1, p. 43). This result may be informally described as follows.

Suppose that a system with Hamiltonian H_0 possesses a symmetry group G. Then, under appropriate hypotheses, the system will possess a corresponding conservation law. This law is embodied in the existence of a vector-valued function $\mathbf{J} : P \to \mathfrak{g}^*$ ($\mathfrak{g}$ denoting the Lie algebra of G), known as a *momentum map*, that is constant on solution curves $t \mapsto x_t$ of X_{H_0}:

$$\frac{d}{dt}\mathbf{J}(x_t) = 0 \ .$$

Next, consider a perturbation to H_0 of the form

$$H = H_0 + \epsilon F \ ,$$

where F is arbitrary. Furthermore, assume that the symmetry is sufficiently large so as to enforce integrability in the unperturbed system[9]. Let $t \mapsto x_t$ be a solution of the *perturbed* Hamiltonian H, and let $\mathcal{O} \subset \mathfrak{g}^*$ denote the co-adjoint orbit through the initial point $\mathbf{J}(x_0) \in \mathfrak{g}^*$. Then we show that there exist positive constants a, b, c, t_0 and r_0, such that for all sufficiently small $\epsilon \geqslant 0$, one has (see Fig. 1)

$$|t| \leqslant t_0 \exp(c\epsilon^{-a}) \quad \Longrightarrow \quad |\mathbf{J}(x_t) - \mathcal{O}| \leqslant r_0 \epsilon^b \ .$$

Here $|\mathbf{J}(x_t) - \mathcal{O}|$ denotes the distance of $\mathbf{J}(x_t)$ from the orbit $\mathcal{O}$, measured using some Ad^*-invariant inner product on $\mathfrak{g}^*$. The estimate holds provided the Hamiltonian is real-analytic and satisfies an appropriate 'convexity' condition. One must also assume that action-group coordinates can be constructed in an appropriate neighborhood of x_0 (see below). The constants appearing in the estimate depend on the unperturbed Hamiltonian H_0, on the magnitude and analyticity properties of F, and on characteristics of G. In directions tangential to the orbits, 'fast' motions, corresponding to those observed by Fassò, are possible.

[9]To be precise, we require that the Marsden-Weinstein reduced spaces be zero-dimensional; see Chap. 3. In some cases the two-dimensional case can be treated also; see Remark 3.9.

Constructing action-group coordinates

The local existence of action-group coordinates has been proven by Dazord and Delzant (1987, Section 5) for suitable non-commutatively integrable systems. We briefly survey this and related work at the end of Chap. 3. In Part 2 of these memoirs we offer an alternative construction that we have applied to several mechanical systems. In this approach the construction of action-group coordinates is reduced to the construction of conventional action-*angle* coordinates in an associated lower dimensional phase space known as a *symplectic cross-section.* This approach may be useful to practitioners of perturbation theory, who are already intimately familiar with conventional action-angle coordinates. The construction may be regarded as a particular application of the so-called 'symplectic cross-section theorem' (Guillemin, Lerman and Sternberg, 1996; Guillemin and Sternberg, 1984).

Action-group coordinates can be constructed *globally* using the cross-section method precisely when action-angle coordinates can be constructed globally in the symplectic cross-section (Theorem 10.2, p. 58). Necessary and sufficient conditions for the existence of global action-angle coordinates in a Hamiltonian system are already known (Duistermaat, 1980). In Chap. 11 we apply the symplectic cross-section technique to the axisymmetric Euler-Poinsot rigid body. Perturbations to this problem have been studied by Benettin and Fassò (1996) using partial action-angle coordinate charts.

Unfortunately to construct action-group coordinates in the neighborhood of a point, one must assume that its image under the momentum map is a *regular* point of the co-adjoint action. In addition, the symmetry group must be acting *freely.* Nevertheless there do exist nontrivial examples for which action-group coordinates can be constructed, as the axisymmetric Euler-Poinsot rigid body demonstrates. Other examples include the problem of geodesics on S^2 (and hence the 'regularized' 2D Kepler problem[10]) and the 1 : 1 resonance. Moreover, it seems likely that techniques such as those outlined here (both the analytic and geometric) will generalize to cases where more sophisticated Hamiltonian G-space normal forms are applicable. We discuss this possibility further in Chap. 13.

[10]See Moser (1970).

Outline of Part 1

Part 1 of this memoir focuses on dynamics. After dispensing in Chap. 1 with some Lie theoretic preliminaries, we begin in Chap. 2 with a résumé of the action-group coordinate framework. In Chap. 3 we state conditions ensuring the existence of action-group coordinates in a given system (Theorem 3.10, p. 19). We include a description of geodesic motions on S^2, where the existence of action-group coordinates is fairly transparent.

Chap. 4 describes the geometry of the unperturbed dynamics of a system in action-group coordinates, as well as the dynamics obtained after naively applying 'multi-phase averaging' to the perturbation. This will explain, from the symmetry point of view, the origin of the 'fast' motions tangent to the co-adjoint orbits, and motivate the Nekhoroshev estimates for the transverse motions (Corollary 7.1, p. 43), which we deduce in Sections 5–7. These estimates follow from Theorem 5.8 (p. 30), an abstract generalization of Nekhoroshev's theorem that we prove in Appendix A using the method of Lochak (1992).

Some limitations, as well as opportunities for further investigation, are discussed in Chap. 13.

Outline of Part 2

Part 2 describes geometric constructions underlying the analyses given in Part 1. Chap. 8 studies a special class of Hamiltonian G-spaces, namely those possessing points of 'regular co-adjoint orbit type.' It is shown (Theorem 8.14, p. 50) that all geometric information describing such a space is encoded in a certain Hamiltonian T-space of lower dimension (its *symplectic cross-section*).

In Chap. 9 we show how the action-group model space can be realized as a symplectic submanifold of T^*G. We derive expressions for the symplectic structure and for Hamiltonian vector fields stated Chap. 2.

In Chap. 10 we use the results of Chap. 8 to reduce the problem of constructing action-group coordinates in an integrable system, to the construction of action-*angle* coordinates in its symplectic cross-section (Theorem 10.2, p. 58). We tackle the problem of *global* existence by relating work of Duistermaat (1980). We also prove 'semi-global' results, which are more convenient to apply in concrete examples. This includes a proof of Theorem 3.10 stated in Chap. 3 (see Corollary 10.12, p. 63). An application to the axisymmetric Euler-Poinsot rigid body is made in Chap. 11.

The constructions of Chap. 10 apply to systems that have zero dimensional Marsden-Weinstein reduced spaces, with respect to some known symmetry group G (we call this *geometric* integrability). In Chap. 12 we show how one can sometimes enlarge G in a system with *two* dimensional reduced spaces (*dynamic* integrability) to $G' \equiv G \times S^1$, in such a way as to render the G'-reduced spaces zero dimensional. An application to rigid bodies is given in Appendix D.

In Appendix E we include a detailed proof of Weinstein's Symplectic Leaf Correspondence Theorem, which we use in Chap. 12.

Part 1

Dynamics

CHAPTER 1

Lie-Theoretic Preliminaries

The purpose of this chapter is to recall some basic Lie theoretic results and to establish some associated terminology and notation.

Weyl chambers

When one replaces the torus $\mathbb{T}^n$ in the action-angle model space $\mathbb{T}^n \times \mathbb{R}^n$ with a compact connected Lie group G, the natural generalization of the action space $\mathbb{R}^n$ turns out to be a *Weyl chamber* of G. We now recall the standard 'geometric' definition of this object.

1.1 Definition. If a group G acts on a manifold X, then an orbit in X is *regular* if there exist no orbits in X of strictly greater dimension. A point $x \in X$ is called *regular* if it lies on a regular orbit. Let $\mathfrak{g}$ denote the Lie algebra of a compact connected Lie group G and let $\mathfrak{t} \subset \mathfrak{g}$ be any maximal Abelian subalgebra. Denote by $\mathfrak{g}_{\text{reg}} \subset \mathfrak{g}$ the regular points of the adjoint action of G, $g \cdot \xi \equiv \mathrm{Ad}_g\, \xi$ $(\xi \in \mathfrak{g})$. A connected component $\mathfrak{t}_0$ of the set $\mathfrak{t} \cap \mathfrak{g}_{\text{reg}}$ is called an (open) *Weyl chamber of G in $\mathfrak{g}$*.

Some related Lie theoretic facts needed in the sequel are summarized below.

1.2 Theorem. *Let G be a compact connected Lie group. Then:*

1. *$\mathfrak{g}_{\text{reg}} \subset \mathfrak{g}$ is open and dense.*
2. *All maximal tori of G are conjugate.*
3. *Every $g \in G$ lies in some maximal torus.*
4. *The Lie algebra $\mathfrak{t}$ of any maximal torus $T \subset G$ is a maximal Abelian subalgebra.*
5. *Every point $\xi \in \mathfrak{g}$ belongs to at least one maximal Abelian subalgebra.*
6. *The map sending a maximal torus to its Lie algebra is a bijection between the maximal tori of G and the maximal Abelian subalgebras of $\mathfrak{g}$. If $\mathfrak{t}$ is a maximal Abelian subalgebra, then $\mathfrak{t} \cap \mathfrak{g}_{\text{reg}} \subset \mathfrak{t}$ is open and dense, and the isotropy group G_ξ is identical for all points $\xi \in \mathfrak{t} \cap \mathfrak{g}_{\text{reg}}$. This group G_ξ is precisely the inverse of $\mathfrak{t}$ under the above correspondence, i.e., G_ξ is the unique maximal torus whose Lie algebra is $\mathfrak{t}$.*
7. *Each regular adjoint orbit intersects each Weyl chamber in exactly one point.*
8. *If $\mathfrak{t}$ is a maximal Abelian subalgebra and we define $\mathfrak{t}^\perp \equiv [\mathfrak{g}, \mathfrak{t}]$, then one has the direct sum decomposition*

$$\mathfrak{g} = \mathfrak{t} \oplus \mathfrak{t}^\perp \ .$$

Proofs of the above facts can be found in, e.g., Bröcker and tom Dieck (1985). The reader may view these facts as generalizations to compact groups of familiar properties of the rotation group SO(3). For example, 1.2.3 corresponds to Euler's theorem that every rotation is a rotation about some fixed axis.

Henceforth G denotes a compact connected Lie group with Lie algebra $\mathfrak{g}$.

Since G is compact, there exists an Ad-invariant inner product $(\xi, \eta) \mapsto \xi \cdot \eta$ on $\mathfrak{g}$. So $\mathrm{Ad}_g \xi \cdot \mathrm{Ad}_g \eta = \xi \cdot \eta$ for all $\xi, \eta \in \mathfrak{g}$ and $g \in G$. Infinitesimally, we have, for arbitrary $\xi, \eta, \zeta \in \mathfrak{g}$,

1.3
$$\mathrm{ad}_\zeta \xi \cdot \eta + \xi \cdot \mathrm{ad}_\zeta \eta = [\zeta, \xi] \cdot \eta + \xi \cdot [\zeta, \eta] = 0 \ .$$

The inner product on $\mathfrak{g}$ has a unique extension to a (non-degenerate) $\mathbb{C}$-bilinear form on its complexification $\mathfrak{g}^{\mathbb{C}} \equiv \mathfrak{g} \otimes_{\mathbb{R}} \mathbb{C}$, which we also denote by $(\xi, \eta) \mapsto \xi \cdot \eta$. Identity 1.3 generalizes to an identity on $\mathfrak{g}^{\mathbb{C}}$.

Weyl chambers in $\mathfrak{g}^*$

Fix a maximal torus T and corresponding maximal Abelian subalgebra $\mathfrak{t}$. Equation 1.3 says that $\mathrm{ad}_\zeta : \mathfrak{g} \to \mathfrak{g}$ is skew-symmetric with respect to the Ad-invariant product. Its image and kernel are therefore orthogonal for any $\zeta \in \mathfrak{t}$. This fact, and the commutativity of $\mathfrak{t}$, easily show that the decomposition 1.2.8 is orthogonal (explaining the notation $\mathfrak{t}^\perp$). If we define $\underline{\mathfrak{t}} \equiv \mathrm{Ann}\, \mathfrak{t}^\perp$ (Ann denotes the annihilator) and $\underline{\mathfrak{t}}^\perp \equiv \mathrm{Ann}\, \mathfrak{t}$, then we obtain the dual decomposition

1.4
$$\mathfrak{g}^* = \underline{\mathfrak{t}} \oplus \underline{\mathfrak{t}}^\perp \ .$$

Let $\varphi : \mathfrak{g} \to \mathfrak{g}^*$ denote the isomorphism induced by the Ad-invariant inner product. Then the orthogonality of the decomposition 1.2.8 implies

1.5
$$\varphi(\mathfrak{t}) = \underline{\mathfrak{t}} \ .$$

The isomorphism φ is G-equivariant if we let G act on $\mathfrak{g}^*$ via the co-adjoint action, $g \cdot \mu \equiv \mathrm{Ad}^*_{g^{-1}} \mu$. It thus establishes an equivalence between the adjoint and co-adjoint representations. In particular, by virtue of 1.5, φ maps a Weyl chamber $\mathfrak{t}_0 \subset \mathfrak{t}$ to a connected component of $\underline{\mathfrak{t}} \cap \mathfrak{g}^*_{\mathrm{reg}}$, which we call a *Weyl chamber in* $\mathfrak{g}^*$. Here $\mathfrak{g}^*_{\mathrm{reg}}$ denotes the regular points of the co-adjoint action. The equivalence of the two representations establishes analogues of 1.2.5, 1.2.6 and 1.2.7 for the co-adjoint action:

1.6 Corollary.

1. *For each $\mu \in \mathfrak{g}^*$ there exists a maximal Abelian subalgebra $\mathfrak{t} \subset \mathfrak{g}$ such that $\mu \in \underline{\mathfrak{t}} \equiv \mathrm{Ann}[\mathfrak{g}, \mathfrak{t}]$.*
2. *Let $\mathcal{W}$ be a Weyl chamber in $\mathfrak{g}^*$, i.e., a connected component of $\underline{\mathfrak{t}} \cap \mathfrak{g}^*_{\mathrm{reg}}$, for some maximal Abelian subalgebra $\mathfrak{t} \subset \mathfrak{g}$. Then $G_\mu = T$ for all $\mu \in \mathcal{W}$, where T is the maximal torus with Lie algebra $\mathfrak{t}$.*
3. *Each regular co-adjoint orbit intersects each Weyl chamber in $\mathfrak{g}^*$ in exactly one point.*

CHAPTER 2

Action-Group Coordinates

In this section we describe the action-group model space. Chap. 3 will state conditions under which this model space is realizable in a particular system, and give a brief historical sketch of its origins.

Let G denote a (real-analytic) compact connected Lie group and let $G^{\mathbb{C}}$ denote its complexification (see, e.g., Bröcker and tom Dieck (1985)). For computing estimates later on we need to assume that G is realized as a real-analytic subgroup of $\mathrm{SO}(n_G, \mathbb{R})$ for some integer n_G. Since G is compact and connected this is always possible, by a corollary of the Peter-Weyl theorem (see, e.g., op. cit, Theorem 4.1 and Exercise 4.7.1). We may then identify the Lie algebra of G with a subalgebra $\mathfrak{g} \subset \mathbb{R}^{n_G \times n_G}$. The adjoint action can be written as $\mathrm{Ad}_g \xi = g\xi g^{-1}$, and the Lie bracket as $[\xi_1, \xi_2] \equiv \mathrm{ad}_{\xi_1} \xi_2 = \xi_1\xi_2 - \xi_2\xi_1$. The complexification $G^{\mathbb{C}}$ of G can be identified with a complex subgroup of $\mathrm{SO}(n_G, \mathbb{C})$. The Lie algebra $\mathfrak{g}^{\mathbb{C}}$ of $G^{\mathbb{C}}$ is identifiable with the complex subalgebra $\mathfrak{g} \oplus i\mathfrak{g} \subset \mathbb{C}^{n_G \times n_G}$.

Henceforth $T \subset G$ denotes a fixed maximal torus, $\mathfrak{t}$ its Lie algebra, and $\mathcal{W} \subset \underline{\mathfrak{t}}$ a Weyl chamber in $\mathfrak{g}^*$.

The model space and its symplectic structure

The natural projection $\mathfrak{g}^* \to \mathfrak{t}^*$ (the dual map of inclusion) has kernel $\underline{\mathfrak{t}}^{\perp}$ and thus restricts, by virtue of 1.4, to an isomorphism $i : \underline{\mathfrak{t}} \to \mathfrak{t}^*$. This map identifies $\mathcal{W}$ with an open set $\mathfrak{t}_0^* \equiv i(\mathcal{W}) \subset \mathfrak{t}^*$. One calls $\mathfrak{t}_0^*$ a Weyl Chamber also.

We define the *action-group model space* for a compact connected Lie group G as $G \times \mathfrak{t}_0^*$. A natural symplectic structure ω_G^* on $G \times \mathfrak{t}_0^*$, generalizing the canonical structure $\sum_j dq_j \wedge dp_j$ on $\mathbb{T}^n \times \mathbb{R}^n$, is given by the following proposition:

2.1 Proposition. *Equip T^*G with its natural symplectic structure. For each $\alpha \in \mathfrak{g}^*$, let α_G denote the left-invariant one-form on G with $\alpha_G(\mathrm{id}_G) = \alpha$ (here viewing one-forms as sections of the cotangent bundle). Then the embedding $G \times \mathfrak{t}_0^* \hookrightarrow \mathrm{T}^*G$ that maps (g, p) to $(i^{-1}(p))_G(g)$, maps $G \times \mathfrak{t}_0^*$ onto a* ***symplectic*** *submanifold of T^*G.*

We define ω_G^* to be the symplectic structure on $G \times \mathfrak{t}_0^*$ pulled back by this embedding. Note that ω_G^* does not depend on the choice of Ad-invariant inner product on $\mathfrak{g}$, or on the realization of G as a linear group. Proposition 2.1 follows from a more general observation we shall make in Part 2, or may be verified directly.

The injection $\varphi^{-1} \circ i^{-1} : \mathfrak{t}^* \hookrightarrow \mathfrak{g}$ maps $\mathfrak{t}^*$ isomorphically onto $\mathfrak{t}$ (see 1.5), and so identifies $\mathfrak{t}_0^*$ with a Weyl chamber $\mathfrak{t}_0 \subset \mathfrak{t}$ in $\mathfrak{g}$. For computing estimates later on, it is convenient for us to make the corresponding identification of $G \times \mathfrak{t}_0^*$ with $G \times \mathfrak{t}_0$. We now describe explicitly the symplectic structure on $G \times \mathfrak{t}_0$ induced by this identification. This symplectic structure of course *does* depend on the choice of Ad-invariant inner product. Since we eventually need a complexification of the model space, we develop notation which also makes sense on the complex manifold $G^{\mathbb{C}} \times \mathfrak{t}^{\mathbb{C}}$ containing $G \times \mathfrak{t}_0$ ($\mathfrak{t}^{\mathbb{C}} \equiv \mathfrak{t} \oplus i\mathfrak{t}$). For $\xi_0 \in \mathfrak{g}^{\mathbb{C}}$, $\tau_0 \in \mathfrak{t}^{\mathbb{C}}$ and $(g,p) \in G^{\mathbb{C}} \times \mathfrak{t}^{\mathbb{C}}$, define the (complex) vector $(\xi_0, \tau_0)_{g,p}$ tangent to $G^{\mathbb{C}} \times \mathfrak{t}^{\mathbb{C}}$ at (g,p) by

2.2
$$(\xi_0, \tau_0)_{g,p} \equiv \frac{d}{dt}(g\exp(t\xi_0), p + t\tau_0)\Big|_{t=0} \ .$$

Note that every tangent vector in $\mathrm{T}_{(g,p)}(G^{\mathbb{C}} \times \mathfrak{t}^{\mathbb{C}})$ is of this form.

On $G^{\mathbb{C}} \times \mathfrak{t}^{\mathbb{C}}$ define the one-form Θ_G by

2.3
$$\langle \Theta_G, (\xi_0, \tau_0)_{g,p} \rangle \equiv p \cdot \xi_0 \ ,$$

and define the two-form ω_G on $G^{\mathbb{C}} \times \mathfrak{t}^{\mathbb{C}}$ by

2.4
$$\omega_G \equiv -d\Theta_G \ .$$

We claim:

> *Restricted to $G \times \mathfrak{t}_0 \subset G^{\mathbb{C}} \times \mathfrak{t}^{\mathbb{C}}$, ω_G agrees with the (real) symplectic structure ω_G^* on $G \times \mathfrak{t}_0^*$ defined above, after making the identification $G \times \mathfrak{t}_0 \cong G \times \mathfrak{t}_0^*$ discussed above.*

For a proof, seee Lemma 9.4, Part 2. The above formulas also appear in Dazord and Delzant (1987, Section 5) (who obtain it via a different route). For our applications to perturbation theory, we need explicit equations of motion and an explicit formula for the Poisson bracket. We turn to these next.

Hamiltonian vector fields in action-group coordinates

If $\xi : G^{\mathbb{C}} \times \mathfrak{t}^{\mathbb{C}} \to \mathfrak{g}^{\mathbb{C}}$ and $\tau : G^{\mathbb{C}} \times \mathfrak{t}^{\mathbb{C}} \to \mathfrak{t}^{\mathbb{C}}$ are arbitrary holomorphic maps, we define vector fields $\xi \cdot \frac{\partial}{\partial g}$ and $\tau \cdot \frac{\partial}{\partial p}$ on $G^{\mathbb{C}} \times \mathfrak{t}^{\mathbb{C}}$ by

$$\left(\xi \cdot \frac{\partial}{\partial g}\right)(g,p) \equiv (\xi(g,p), 0)_{g,p} \ , \qquad \left(\tau \cdot \frac{\partial}{\partial p}\right)(g,p) \equiv (0, \tau(g,p))_{g,p} \ .$$

An arbitrary vector field on $G^{\mathbb{C}} \times \mathfrak{t}^{\mathbb{C}}$ is then of the form $\xi \cdot \frac{\partial}{\partial g} + \tau \cdot \frac{\partial}{\partial p}$ for some vector-valued functions $\xi : G^{\mathbb{C}} \times \mathfrak{t}^{\mathbb{C}} \to \mathfrak{g}^{\mathbb{C}}$ and $\tau : G^{\mathbb{C}} \times \mathfrak{t}^{\mathbb{C}} \to \mathfrak{t}^{\mathbb{C}}$.

Write $\mathfrak{t}^{\perp\mathbb{C}} = \mathfrak{t}^{\perp} \oplus i\mathfrak{t}^{\perp}$ and define $\mathfrak{t}_0^{\mathbb{C}}$ to be the connected component containing $\mathfrak{t}_0$ of the set

$$\{p \in \mathfrak{t}^{\mathbb{C}} \mid \mathrm{ad}_p : \mathfrak{t}^{\perp\mathbb{C}} \to \mathfrak{t}^{\perp\mathbb{C}} \text{ is invertible }\} \ .$$

For $p \in \mathfrak{t}_0^{\mathbb{C}}$, let $\lambda_p : \mathfrak{t}^{\perp\mathbb{C}} \to \mathfrak{t}^{\perp\mathbb{C}}$ denote the inverse of $\mathrm{ad}_p : \mathfrak{t}^{\perp\mathbb{C}} \to \mathfrak{t}^{\perp\mathbb{C}}$. In particular, let us record that

2.5
$$\lambda_p([p,\xi]) = \xi = [p, \lambda_p(\xi)] \qquad (\xi \in \mathfrak{t}^{\perp\mathbb{C}}, p \in \mathfrak{t}_0^{\mathbb{C}})$$

and

2.6
$$\mathfrak{t}_0 = \mathfrak{t} \cap \mathfrak{t}_0^{\mathbb{C}} \ .$$

It is possible to show that the restriction of the (complex) form ω_G to $G^{\mathbb{C}} \times \mathfrak{t}_0^{\mathbb{C}}$ is non-degenerate. If $H : G^{\mathbb{C}} \times \mathfrak{t}_0^{\mathbb{C}} \to \mathbb{C}$ is holomorphic, then the corresponding (complex) *Hamiltonian vector field* X_H is defined by $X_H \lrcorner\, \omega_G = dH$. Indeed

2.7
$$X_H = \xi_H \cdot \frac{\partial}{\partial g} + \tau_H \cdot \frac{\partial}{\partial p}$$

where

$$\xi_H(g,p) \equiv \frac{\partial H}{\partial p}(g,p) + \lambda_p \sigma^\perp \frac{\partial H}{\partial g}(g,p) \tag{2.8}$$

and

$$\tau_H(g,p) \equiv -\sigma \frac{\partial H}{\partial g}(g,p) \ . \tag{2.9}$$

Here $\sigma : \mathfrak{g} \to \mathfrak{t}$ denotes the projection along $\mathfrak{t}^\perp$, and $\sigma^\perp : \mathfrak{g} \to \mathfrak{t}^\perp$ that along $\mathfrak{t}$. These formulas are the complexified version of formulas on $G \times \mathfrak{t}_0 \cong G \times \mathfrak{t}_0^*$ we will derive as Proposition 9.13, Part 2. The vector valued functions $\frac{\partial H}{\partial g} : G^{\mathbb{C}} \times \mathfrak{t}_0^{\mathbb{C}} \to \mathfrak{g}^{\mathbb{C}}$ and $\frac{\partial H}{\partial p} : G^{\mathbb{C}} \times \mathfrak{t}_0^{\mathbb{C}} \to \mathfrak{t}^{\mathbb{C}}$ appearing in 2.8 and 2.9 are defined implicitly by

$$\frac{\partial H}{\partial g}(g,p) \cdot \xi_0 = \langle dH, (\xi_0, 0)_{g,p} \rangle = \frac{d}{dt} H(g \exp(t\xi_0), p)\big|_{t=0} \qquad (\xi_0 \in \mathfrak{g}^{\mathbb{C}})$$

and

$$\frac{\partial H}{\partial p}(g,p) \cdot \tau_0 = \langle dH, (0, \tau_0)_{g,p} \rangle = \frac{d}{dt} H(g, p + t\tau_0)\big|_{t=0} \qquad (\tau_0 \in \mathfrak{t}^{\mathbb{C}}) \ .$$

Here a dot denotes the non-degenerate $\mathbb{C}$-bilinear form on $\mathfrak{g}^{\mathbb{C}}$.

Equations of motion in action-group coordinates

Recalling that we are identifying G with a linear group, we may think of elements of G as matrices and use formulas 2.7–2.9 to write the 'equations of motion' corresponding to a Hamiltonian H as

$$\begin{aligned} \dot g &= g \frac{\partial H}{\partial p} + g \lambda_p \sigma^\perp \frac{\partial H}{\partial g} \\ \dot p &= -\sigma \frac{\partial H}{\partial g} \end{aligned} \qquad \left((g,p) \in G^{\mathbb{C}} \times \mathfrak{t}_0^{\mathbb{C}} \right) \ . \tag{2.10}$$

Although one can (abstractly) make sense of these equations without the realization of G as a linear group, our estimates later on will depend on the vector space structure of $\mathbb{C}^{n_G \times n_G}$ in which we have assumed $G^{\mathbb{C}}$ to be embedded.

2.11 Remark. It is worth comparing the above equations to those corresponding to Hamiltonian vector fields on $T^*G \cong G \times \mathfrak{g}^*$ (left trivialization, say). These may be derived from the formula for the symplectic structure in Abraham and Marsden (1978, Proposition 4.4.1). In the present notation, and identifying $\mathfrak{g}^*$ with $\mathfrak{g}$ as above, these equations take the form

$$\begin{aligned} \dot g &= g \frac{\partial H}{\partial \mu} \\ \dot \mu &= -\frac{\partial H}{\partial g} + \left[\mu, \frac{\partial H}{\partial \mu} \right] \end{aligned} \qquad \left((g,\mu) \in G \times \mathfrak{g} \right) \ .$$

A Hamiltonian invariant with respect to the cotangent lifted left action of G (see p. 53) is represented by a function $H_0(g,\mu) = h(\mu)$. However, as we will explain in Remark 4.6, the 'coordinates' $G \times \mathfrak{g}$ are poorly suited for an analysis of perturbations to general Hamiltonians of this form (unless G is Abelian).

2.12 Example.
Take $G = S^1$. Then G (resp. $G^{\mathbb{C}}$) is realized as SO(2) (resp. SO(2, $\mathbb{C}$)) and we have $\mathfrak{t}_0^{\mathbb{C}} = \mathfrak{g}^{\mathbb{C}}$, $\mathfrak{t}^{\perp} = 0$. Coordinates $(g, p) \in G^{\mathbb{C}} \times \mathfrak{t}_0^{\mathbb{C}}$ are of the form

$$g = \begin{bmatrix} \cos\theta & -\sin\theta \\ \sin\theta & \cos\theta \end{bmatrix} , \quad \theta \in \mathbb{C} \ ; \qquad p = \begin{bmatrix} 0 & -I \\ I & 0 \end{bmatrix} , \quad I \in \mathbb{C} \ .$$

For an inner product on $\mathfrak{g}$, one chooses $p_1 \cdot p_2 \equiv I_1 I_2$. If one writes $H'(\theta, I) \equiv H(g, p)$, then 2.10 takes the form

$$\dot{\theta} = \frac{\partial H'}{\partial I} , \qquad \dot{I} = -\frac{\partial H'}{\partial \theta} ,$$

which are the familiar form of Hamilton's equations.

One generalizes the above argument to $G = \mathbb{T}^n \cong \mathrm{SO}(2) \times \cdots \times \mathrm{SO}(2)$ by realizing G as the set of block diagonal $2n \times 2n$ matrices whose 2×2 blocks are of the form just described for $n = 1$.

The next example is relevant to the problem of geodesic motions on S^2 (see 3.4).

2.13 Example. Take $G = \mathrm{SO}(3)$. Then $\mathfrak{g} = \mathfrak{so}(3)$ can be identified with $\mathbb{R}^3$ via the isomorphism $\xi \mapsto \hat{\xi} : \mathbb{R}^3 \to \mathfrak{so}(3)$ defined by $\hat{\xi} u = \xi \times u$ ($u \in \mathbb{R}^3$). This isomorphism extends uniquely to a $\mathbb{C}$-linear isomorphism $\mathbb{C}^3 \to \mathfrak{so}(3, \mathbb{C})$. Let $\{e_1, e_2, e_3\}$ denote the standard basis of $\mathbb{R}^3$. Choose $T \subset \mathrm{SO}(3)$ to be the rotations about the e_3 axis, so that $\mathfrak{t} = \mathrm{span}\{e_3\} \cong \mathbb{R}$. Then $\mathfrak{t}^{\perp} = \mathrm{span}\{e_1, e_2\} \cong \mathbb{R}^2$. The adjoint action is given simply by $g \cdot \xi = g\xi$ ($g \in \mathrm{SO}(3), \xi \in \mathbb{R}^3$), so that the regular adjoint orbits are the spheres centered at the origin of positive radius. Whence $\mathfrak{g}_{\mathrm{reg}} = \mathbb{R}^3 \backslash \{0\}$. The standard inner product $a \cdot b \equiv a_1 b_1 + a_2 b_2 + a_3 b_3$ is Ad-invariant. A connected component $\mathfrak{t}_0$ of $\mathfrak{t} \cap \mathfrak{g}_{\mathrm{reg}}$ is given by $\mathfrak{t}_0 \equiv \{t e_3 \mid t > 0\} \cong (0, \infty)$. So $G \times \mathfrak{t}_0 \cong \mathrm{SO}(3) \times (0, \infty)$.

One has $\mathfrak{g}^{\mathbb{C}} \cong \mathbb{C}^3$ with $\mathrm{ad}_{\xi} : \mathfrak{g}^{\mathbb{C}} \to \mathfrak{g}^{\mathbb{C}}$ given by $\mathrm{ad}_{\xi}\, \eta = \xi \times \eta$. Fixing $p \in \mathfrak{t}^{\mathbb{C}} \cong \mathbb{C}$, one finds that $\mathrm{ad}_p : \mathfrak{t}^{\perp\mathbb{C}} \to \mathfrak{t}^{\perp\mathbb{C}}$ ($\mathfrak{t}^{\perp\mathbb{C}} \cong \mathbb{C}^2$) is given by $\mathrm{ad}_p(\xi_1, \xi_2) = (-p\xi_2, p\xi_1)$. Therefore, $\mathfrak{t}_0^{\mathbb{C}} \cong \mathbb{C}\backslash\{0\}$ and $\lambda_p : \mathfrak{t}^{\perp\mathbb{C}} \to \mathfrak{t}^{\perp\mathbb{C}}$ is given by $\lambda_p(\xi_1, \xi_2) = (\xi_2/p, -\xi_1/p)$ ($p \in \mathbb{C}\backslash\{0\}$). The projections $\sigma : \mathfrak{g}^{\mathbb{C}} \to \mathfrak{t}^{\mathbb{C}}$ and $\sigma^{\perp} : \mathfrak{g}^{\mathbb{C}} \to \mathfrak{t}^{\perp\mathbb{C}}$ are given by $\sigma(\xi_1, \xi_2, \xi_3) = \xi_3$, $\sigma^{\perp}(\xi_1, \xi_2, \xi_3) = (\xi_1, \xi_2)$.

As the reader will readily verify, the equations of motion 2.10 take the form of (*) on p. xvii.

The Poisson bracket in action-group coordinates

Our convention for defining Poisson brackets is $\{u, v\} \equiv X_v \lrcorner X_u \lrcorner \omega$. If $t \mapsto x_t$ is an integral curve of a Hamiltonian vector field X_H then, according to this definition,

$$\frac{d}{dt} u(x_t) = \{u, H\}(x_t)$$

for any function u.

2.14 Lemma. *The Poisson bracket on $G^{\mathbb{C}} \times \mathfrak{t}_0^{\mathbb{C}}$ is given by*

$$\begin{aligned} \{u, v\}(g, p) = {} & \frac{\partial u}{\partial g}(g, p) \cdot \frac{\partial v}{\partial p}(g, p) - \frac{\partial v}{\partial g}(g, p) \cdot \frac{\partial u}{\partial p}(g, p) \\ & - p \cdot [\lambda_p \sigma^{\perp} \frac{\partial u}{\partial g}(g, p), \lambda_p \sigma^{\perp} \frac{\partial v}{\partial g}(g, p)] \ . \end{aligned}$$

PROOF.

$$\begin{aligned}\{u,v\}(g,p) &=< du, X_v(g,p) > \\ &= \xi_v(g,p)\cdot\frac{\partial u}{\partial g}(g,p)+\tau_v(g,p)\cdot\frac{\partial u}{\partial p}(g,p) \qquad \text{by 2.7}\\ &= (\frac{\partial v}{\partial p}(g,p)+\lambda_p\sigma^{\perp}\frac{\partial v}{\partial g}(g,p))\cdot\frac{\partial u}{\partial g}(g,p)-\sigma\frac{\partial v}{\partial g}(g,p)\cdot\frac{\partial u}{\partial p}(g,p)\end{aligned}$$

by 2.8 and 2.9,

$$\begin{aligned}&= \frac{\partial u}{\partial g}(g,p)\cdot\frac{\partial v}{\partial p}(g,p)-\frac{\partial v}{\partial g}(g,p)\cdot\frac{\partial u}{\partial p}(g,p)\\ &+\lambda_p\sigma^{\perp}\frac{\partial v}{\partial g}(g,p)\cdot\frac{\partial u}{\partial g}(g,p) \ .\end{aligned}$$

The last term can be written

$$\begin{aligned}\lambda_p\sigma^{\perp}\frac{\partial v}{\partial g}(g,p)\cdot\frac{\partial u}{\partial g}(g,p) &= \lambda_p\sigma^{\perp}\frac{\partial v}{\partial g}(g,p)\cdot\sigma^{\perp}\frac{\partial u}{\partial g}(g,p)\\ &= \lambda_p\sigma^{\perp}\frac{\partial v}{\partial g}(g,p)\cdot[p,\lambda_p\sigma^{\perp}\frac{\partial u}{\partial g}(g,p)] \qquad \text{by 2.5}\\ &= -p\cdot[\lambda_p\sigma^{\perp}\frac{\partial u}{\partial g}(g,p),\lambda_p\sigma^{\perp}\frac{\partial v}{\partial g}(g,p)] \qquad \text{by 1.3} \ .\end{aligned}$$

□

CHAPTER 3

On the Existence of Action-Group Coordinates

In this chapter we formulate the main theorem on the existence of action-group coordinates for a Hamiltonian system with a non-Abelian symmetry. A proof of the theorem is postponed to Part 2, where the construction of the coordinates is reduced to the construction of conventional action-angle coordinates in an associated lower-dimensional phase space.

Hamiltonian system is said to possess a *symmetry* when there exists a Lie group G acting on the system's phase space (P, ω), with respect to which the Hamiltonian function H_0 is invariant: $H_0(g \cdot x) = H_0(x)$ for all $g \in G$ and $x \in P$. The group G is said to be acting in a *Hamiltonian fashion* if it acts by symplectic diffeomorphisms, and if the infinitesimal generators ξ_P ($\xi \in \mathfrak{g}$) of the action are (global) Hamiltonian vector fields on P. In that case there exists a map $\mathbf{J} : P \to \mathfrak{g}^*$, called a *momentum map,* with the property that it delivers Hamiltonian functions $J_\xi : P \to \mathbb{R}$ for the generators according to the formula $J_\xi(x) \equiv \langle \mathbf{J}(x), \xi \rangle$. Here $\langle \cdot \, , \cdot \rangle$ denotes the natural pairing between $\mathfrak{g}^*$ and $\mathfrak{g}$. Thus $\xi_P = X_{J_\xi}$, where X_f denotes the Hamiltonian vector field corresponding to a function f, i.e., the vector field defined through $X_f \lrcorner\, \omega = df$.

Noether's theorem states that the functions J_ξ (which Poisson-commute only in the Abelian case) are integrals of motion for the G-invariant Hamiltonian H_0. In particular, $\mathbf{J}(x_t) \in \mathfrak{g}^*$ is constant for all solution curves $t \mapsto x_t$.

See Marsden and Ratiu (1994), Abraham and Marsden (1978, Chapter 4) or Guillemin and Sternberg (1984) for background on momentum maps and for an introduction to the geometric point of view of symmetry in mechanics. For a rapid introduction and recent survey, see Marsden (1992).

System symmetry breaking

Let (P, ω) be a symplectic manifold and let G be a compact connected Lie group acting on P in a Hamiltonian fashion, with momentum map $\mathbf{J} : P \to \mathfrak{g}^*$. We suppose that $\mathbf{J}$ is G-equivariant, with G acting on $\mathfrak{g}^*$ via the co-adjoint action. The quadruple $(P, \omega, G, \mathbf{J})$ (or simply P) will then be called a *Hamiltonian G-space.*

3.1 Example (Action-angle coordinates). Take $P = \mathbb{T}^n \times \mathbb{R}^n$ and $\omega = \Sigma dq_j \wedge dp_j$. Let $G = \mathbb{T}^n \equiv \mathbb{R}^n / 2\pi\mathbb{Z}^n$ act on P according to $\theta \cdot (q, p) \equiv (\theta + q, p)$. This action is Hamiltonian with equivariant momentum map $\mathbf{J} : P \to \mathfrak{g}^* \cong \mathbb{R}^n$ given by $\mathbf{J}(q, p) = p$. A Hamiltonian is G-invariant precisely when it depends only on the action coordinates p.

A basic problem of perturbation theory is to understand Hamiltonians $H : P \to \mathbb{R}$ with 'broken symmetry,' namely those of the form

$$H = H_0 + F \ ,$$

where H_0 is G-invariant and F is suitable small.

Orbit type conditions

If a group G acts on a space M, then we denote the point stabilizer (isotropy) subgroup at $x \in M$ by G_x. Recall that the *orbit type* of x is the conjagacy class of subgroups of G with representative G_x. It is denoted (G_x).

Reconsider the case of a Hamiltonian G-space as above. In addition to the orbit type of a point $x \in P$ we shall also, in a slight abuse of terminology, refer to its *co-adjoint orbit type*. This is defined to be the orbit type of $\mathbf{J}(x) \in \mathfrak{g}^*$.

It follows from a general fact about compact group actions (see, e.g., Bredon (1972)) that if P is connected then there exists an open dense subset of P whose points are of uniform orbit type (called the *maximal* orbit type). From the point of view of understanding the fate of generic initial conditions in the symmetry breaking problem above, it is reasonable to restrict attention to this open G-invariant subspace. In all applications of which this author is aware, one can also find an open dense subset of P with constant *co-adjoint* orbit type. (Whether this holds in general is not immediately clear[1].) Such a subset is also G-invariant, by momentum map equivariance.

The preceding arguments suggest that, from the point of view of perturbation theory, an interesting class of Hamiltonian G-spaces is those spaces with simultaneously a constant orbit and co-adjoint orbit type. The simplest case is a space in which G acts freely and $\mathbf{J}(P) \subset \mathfrak{g}^*_{\mathrm{reg}}$. In this case all points in P have orbit type (id_G) and (by 1.6.1, 1.6.2 and 1.2.2) co-adjoint orbit type (T), where T is a maximal torus of G. These are the spaces we shall consider here and are precisely those spaces for which action-group coordinates can be constructed, under an appropriate integrability hypothesis. Note that if G is Abelian, then $\mathfrak{g}^*_{\mathrm{reg}} = \mathfrak{g}^*$, so that the condition $\mathbf{J}(P) \subset \mathfrak{g}^*_{\mathrm{reg}}$ is always satisfied in that case.

3.2 Example (Action-group coordinates). Take $P \equiv G \times \mathfrak{t}_0$ and $\omega \equiv \omega_G$ (notation as in Chap. 2). The group G acts freely on P according to $g \cdot (h,p) \equiv (gh,p)$. We claim that this action admits an equivariant momentum map.

Writing $\Phi_g(h,p) \equiv g \cdot (h,p)$, one can deduce from 2.3 that Θ_G is invariant under pull-back by Φ_g, for all $g \in G$. As a consequence, the Lie derivative of Θ_G along any infinitesimal generator ξ_P $(\xi \in \mathfrak{g})$ vanishes. Therefore, by Cartan's 'magic formula' and 2.4, one has

$$\xi_P \lrcorner\, \omega_G = d(\xi_P \lrcorner\, \Theta_G) \qquad (\xi \in \mathfrak{g}) \ .$$

Thus the generator ξ_P is the Hamiltonian vector field corresponding to the function $J^G_\xi \equiv \xi_P \lrcorner\, \Theta_G$. In the notation of 2.2, these generators are given $\xi_P(g,p) = (\mathrm{Ad}_{g^{-1}} \xi, 0)_{g,p}$. One therefore computes

$$J^G_\xi(g,p) = p \cdot (\mathrm{Ad}_{g^{-1}} \xi) = (\mathrm{Ad}_g\, p) \cdot \xi = \langle \varphi^{-1}(\mathrm{Ad}_g\, p), \xi \rangle \ .$$

Whence a momentum map $\mathbf{J}^G : P \to \mathfrak{g}^*$ is given by

3.3
$$\mathbf{J}^G(g,p) = \varphi^{-1}(\mathrm{Ad}_g\, p) = \mathrm{Ad}^*_{g^{-1}} \varphi^{-1}(p) \ .$$

Recall (see p. 9) that φ maps the Weyl chamber $\mathcal{W}$ in $\mathfrak{g}^*$ bijectively onto the chamber $\mathfrak{t}_0$, so that $\varphi^{-1}(p) \in \mathcal{W} \subset \underline{\mathfrak{t}} \cap \mathfrak{g}^*_{\mathrm{reg}}$ for all $p \in \mathfrak{t}_0$. In particular, $\mathbf{J}^G(P) \subset$

---

[1]To settle the question seems to require a detailed understanding of the structure of the momentum map image. So far only the case of *compact* P seems sufficiently well-understood; see Kirwan (1984) for a treatment of the compact case and see Hilgert, Neeb and Plank (1994) for partial results in the non-compact case.

$\mathfrak{g}^*_{\rm reg}$, so that $(G \times \mathfrak{t}_0, \omega_G, G, \mathbf{J}^G)$ is a Hamiltonian G-space satisfying our orbit type conditions.

Of course Example 3.1 is just Example 3.2 in the special case $G = \mathbb{T}^n$.

3.4 Example (Geodesic motions on S^2). Consider a point mass M constrained to move on the surface of a smooth sphere of radius L. One identifies the position of M with a point on the unit sphere S^2, so that the phase space is T^*S^2, equipped with its standard symplectic structure. We identify this space with

$$\mathrm{T}S^2 \cong \{(q, v) \in \mathbb{R}^3 \times \mathbb{R}^3 \mid \|q\| = 1, q \cdot v = 0\} ,$$

using the standard metric on S^2, in which case the symplectic structure is $\omega = -d\Theta$, where

$$\left\langle \Theta, \frac{d}{dt}(q_t, v_t) \Big|_{t=0} \right\rangle \equiv v_0 \cdot \dot{q}_0 .$$

Here $\dot{q}_0 = d/dt|_{t=0}\, q_t$ is to be viewed as an element of $\mathbb{R}^3$. The Hamiltonian is $H_0(q, v) = L^2\|v\|^2/(2M)$, which is G-invariant with respect to the action of $G \equiv \mathrm{SO}(3)$ defined by $g \cdot (q, v) \equiv (gq, gv)$. This action is free and has an equivariant momentum map $\mathbf{J} : P \to \mathfrak{g}^* \cong \mathbb{R}^3$ given by $\mathbf{J}(q, v) \equiv q \times v$. We restrict attention to the open dense G-invariant subset of phase space obtained by removing the zero section:

$$P \equiv \{(q, v) \in \mathrm{T}S^2 \mid v \neq 0\} .$$

Then $\mathbf{J}(P) = \mathbb{R}^3 \backslash \{0\} = \mathfrak{g}^*_{\rm reg}$, so that $(P, \omega, G, \mathbf{J})$ is a Hamiltonian G-space satisfying our orbit type conditions.

3.5 Remark. Bearing in mind Poincaré's model of SO(3),

$$\mathrm{SO}(3) \cong \{(q, v) \in \mathrm{T}S^2 \mid \|v\| = 1\} ,$$

one observes in Example 3.4, that P can be identified with $\mathrm{SO}(3) \times (0, \infty)$, the action-group model space for $G = \mathrm{SO}(3)$ (see 2.13). In fact this identification is *symplectic*[2]. Whence it gives a straightforward realization of action-group coordinates for the problem. Incidentally, in these coordinates the Hamiltonian is given by $H_0(g, p) = L^2p^2/(2M)$, $(g, p) \in \mathrm{SO}(3) \times (0, \infty)$. Also note that neither conventional action-angle coordinates, nor partial action-angle coordinates, can be constructed globally on P in this problem.

Another simple example satisfying our orbit type conditions is the $1:1$ resonance, where $P \equiv \mathbb{R}^4 \backslash \{0\}$ and $G = \mathrm{SU}(2)$. A more complicated example is the axisymmetric Euler-Poinsot rigid body discussed in Chap. 11 ($G = \mathrm{SO}(3) \times S^1$). In both cases one restricts to some open subset of the phase space to ensure the above orbit type conditions.

In more complicated examples one or both of the above conditions on the orbit types often fail. For example, in the problem of geodesic motions on S^3 (see Chap. 13) the symmetry group SO(4) fails to act freely (although most points do have co-adjoint orbit type (T), $T \subset \mathrm{SO}(4)$ a maximal torus). In the 1:1:1 resonance ($P = \mathbb{R}^6 \equiv \mathbb{C}^3$, $G = \mathrm{SU}(3)$; see Example 8.4, p. 47), all non-zero points have orbit type (SU(2)) and co-adjoint orbit (SU(2)), so that both orbit type conditions fail.

These comments not withstanding, the above orbit type conditions are the natural ones to study first.

[2] For a proof, see Part 2.

Integrability

In the case that G acts freely, a reduction in dimension of a Hamiltonian system with symmetry is achieved by *Marsden-Weinstein reduction*:

3.6 Theorem (Marsden and Weinstein (1974), Meyer (1973)).
Let $(P, \omega, G, \mathbf{J})$ be a Hamiltonian G-space and assume G acts freely and properly. Then for any $\mu \in \mathbf{J}(P)$, $\mathbf{J}^{-1}(\mu)$ is a G_μ-invariant submanifold of P and $P_\mu \equiv \mathbf{J}^{-1}(\mu)/G_\mu$ admits the structure of a smooth manifold, with respect to which the natural projection $\gamma_\mu : \mathbf{J}^{-1}(\mu) \to P_\mu$ is a surjective submersion. There is a unique symplectic form ω_μ on P_μ with the property that $\gamma_\mu^* \omega_\mu = i_\mu^* \omega$, where $i_\mu : \mathbf{J}^{-1}(\mu) \to P$ is the inclusion.

The symplectic manifolds P_μ ($\mu \in \mathbf{J}(P)$) are called the *reduced spaces*. Of course, as we assume G is compact, all actions of G are proper.

If H_0 is a G-invariant Hamiltonian, then for each μ in Theorem 3.6, $\mathbf{J}^{-1}(\mu) \subset P$ is an invariant submanifold (by Noether), and there exists a Hamiltonian $H^\mu : P_\mu \to \mathbb{R}$ such that the vector fields $X_{H_0}|\mathbf{J}^{-1}(\mu)$ and X_{H^μ} (defined by $X_{H^\mu} \lrcorner\, \omega_\mu = dH^\mu$) are γ_μ-related. It can be shown that if the integral curves in the reduced space P_μ of X_{H^μ} are known, then the integral curves of X_{H_0} lying in $\mathbf{J}^{-1}(\mu) \subset P$ can be reconstructed by solving *linear* ordinary differential equations with time dependent coefficients (Marsden, Montgomery and Ratiu, 1990). If $\mathbf{J}(P) \subset \mathfrak{g}^*_{\mathrm{reg}}$ then the reduced spaces all have dimension

$$d \equiv \dim P - (\dim G + \operatorname{rank} G) \ .$$

Here $\operatorname{rank} G$ denotes the dimension of the maximal tori of G.

If $d = 0$ or $d = 2$ then Hamilton's equations on the reduced spaces can always be 'solved': In the $d = 0$ case integral curves of X_{H^μ} are trivial (and the reconstruction equations are in fact *autonomous*); in the $d = 2$ case integral curves of X_{H^μ} coincide with level sets of the reduced Hamiltonian H^μ. It is natural to call the full system *integrable* in either case:

3.7 Definition. A Hamiltonian G-space $(P, \omega, G, \mathbf{J})$, with G acting freely and $\mathbf{J}(P) \subset \mathfrak{g}^*_{\mathrm{reg}}$, will be called *geometrically integrable* if $d = 0$, and *dynamically integrable* if $d = 2$. In either case we say that the integrability is *non-commutative* if G is non-Abelian.

For us the more natural notion will be *geometric* integrability; see 3.10 below.

3.8 Examples. The problem of geodesic motions on S^2 (Example 3.4) is geometrically integrable. Action-group coordinates (Example 3.2) also constitute a geometrically integrable space.

3.9 Remark. Restricting attention to the geometrically integrable case does not rule out the treatment of dynamically integrable systems. Under certain conditions one can enlarge the action of G, in a dynamically integrable space, to an action of $G \times S^1$, such that the space becomes geometrically integrable, with respect to the larger group. We discuss this construction in detail in Chap. 12, and apply it to the Euler-Poinsot rigid body in Appendix D.

The existence theorem

The following result states conditions under which action-group coordinates can be constructed in a Hamiltonian G-space. Recall that $\varphi : \mathfrak{g} \to \mathfrak{g}^*$ denotes the isomorphism corresponding to the fixed Ad-invariant inner product on $\mathfrak{g}$.

3.10 Theorem. *Let G be a compact connected Lie group, and let $(P, \omega, G, \mathbf{J})$ be a Hamiltonian G-space on which G is acting freely, and for which $\mathbf{J}(P) \subset \mathfrak{g}^*_{\mathrm{reg}}$. Assume the space is goemetrically integrable in the sense of 3.7. Assume that P is connected, and that each fiber of $\mathbf{J} : P \to \mathbf{J}(P)$ is compact or has a finite number of connected components. Assume that $U \equiv \varphi^{-1}(\mathbf{J}(P)) \cap \mathfrak{t}_0$ (which is open in $\mathfrak{t}_0$) is smoothly contractible. Let G act on $G \times U \subset G \times \mathfrak{t}_0$ as in Example 3.2. Then there exists an equivariant symplectic diffeomorphism $\phi : G \times U \to P$ such that $\mathbf{J} \circ \phi = \mathbf{J}^G$, where $\mathbf{J}^G : G \times U \to \mathfrak{g}^*$ is the momentum map defined by 3.3.*

This theorem holds in either the C^∞ or real-analytic categories. A proof will be given in Part 2 (see Corollary 10.12, p. 49).

3.11 Remark. The contractability hypothesis can be weakened if the fibres of $\mathbf{J}$ are known to be connected (see Chap. 10). On the other hand, if U in 3.10 fails to be contractible, then one may always work 'locally' by replacing P by a connected component of $\mathbf{J}^{-1}(G \cdot V)$, for some appropriate contractible set $V \subset U$. This suffices for the applications to perturbation theory in the sequel.

3.12 Remark. Let $(P, \omega, G, \mathbf{J})$ be a Hamiltonian G-space on which G is acting freely and for which $\mathbf{J}(P) \subset \mathfrak{g}^*_{\mathrm{reg}}$. In that case G_μ is Abelian (a maximal torus in fact) for all $\mu \in \mathbf{J}(P)$. In Part 2 (see Lemma 8.12, p. 49) we shall recall the following formula:

$$\omega(\xi_P(x), \eta_P(x)) = \langle \mathbf{J}(x), [\xi, \eta] \rangle \qquad (x \in P,\ \xi, \eta \in \mathfrak{g}) \ .$$

Since G_μ is Abelian ($\mu \in \mathbf{J}(P)$), the restriction of ω to G_μ-orbits vanishes. It follows that the distribution D on P defined by $D(x) \equiv (\mathrm{T}_x(G_{\mathbf{J}(x)} \cdot x))^\omega$ is coisotropic (see p. 1 for the definition).

Assume P is G-integrable. Then $\dim \mathbf{J}^{-1}(\mu)/G_\mu = 0$ for any $\mu \in \mathbf{J}(P)$. In that case $\ker \mathrm{T}_x\mathbf{J} = \mathrm{T}_x(G_\mu \cdot x)$. Therefore $D = (\ker \mathrm{T}\mathbf{J})^\omega$. But from the definition of a momentum map, $\mathfrak{g}^* \xleftarrow{\mathbf{J}} P \xrightarrow{\rho} P/G$ has the 'dual pair' property (ρ denoting the natural projection), i.e. $\ker \mathrm{T}\mathbf{J}$ and $\ker \mathrm{T}\rho$ are ω-orthogonal distributions. In particular, $D = \ker \mathrm{T}\rho$, i.e., D is the distribution tangent to the foliation $\mathcal{F}$ of P by G-orbits. This foliation, as we have just demonstrated, is coisotropic. Furthermore, as $D^\omega = \ker \mathrm{T}\mathbf{J}$, $\mathcal{F}$ is also symplectically complete. A G-invariant Hamiltonian $H_0 : P \to \mathbb{R}$ is therefore integrable in the more general sense given in the Introduction (p. 1).

Historical remarks

Integrability, in the more general sense given in the Introduction, has its origins in the work of Alan Weinstein and Paulette Libermann. There are two complementary approaches to studying the geometric structure of a coisotropic, symplectically complete foliation $\mathcal{F}$. One may focus either on neighborhoods of the leaves of $\mathcal{F}$, or on neighborhoods of leaves of the symplectic orthogonal foliation $\mathcal{F}^\omega$ (which is *isotropic*[3]). The latter problem, as we have described in the Introduction, was

[3]Meaning that the tangent distribution of $\mathcal{F}^\omega$ is contained by its symplectic orthogonal.

solved by Nekhoroshev (1972) (under the hypothesis of compact leaves), and clarified later in Dazord and Delzant (1987). Attempts to address the former problem, which are more recent, begin with independent studies of Gotay (1982) and Marle (1982; 1983b), which classify the neighborhood of an arbitrary *isolated* coisotropic submanifold (not necessarily a leaf of a *symplectically complete* foliation). The complementary problem of classifying the neighborhood of an isolated isotropic submanifold, and the 'overlapping' problem of classifying the neighborhood of a Lagrangian[4] submanifold, were already solved in Weinstein (1981) and Weinstein (1971) respectively.

In the studies of Gotay and Marle, it is shown that the neighborhood of a coisotropic submanifold $M \subset P$ looks like a neighborhood of the zero section of the dual E^* of its characteristic bundle E, equipped with an appropriate symplectic structure. The *characteristic bundle* of M is defined by

$$E \equiv \{v \in \mathrm{T}M \mid \omega(v, w) = 0 \ \forall w \in \mathrm{T}M\}.$$

In the special case of symmetry considered in Theorem 3.10, $E(x) = \mathrm{T}_x(G_\mu \cdot x)$, where $\mu \equiv \mathbf{J}(x)$.

The symplectic diffeomorphism realizing a neighborhood of M in P as a neighborhood of the zero section of E^*, and the associated symplectic structure bestowed on E^*, both depend on a choice of splitting $\mathrm{T}M = E \oplus V$. However, up to neighborhood equivalences, all choices lead to the same answer.

In Marle (1983a) the case where M is the orbit of a free and geometrically integrable Hamiltonian action is studied. Here it is pointed out that the choice of splitting $\mathrm{T}M = E \oplus V$ can be reduced to the choice of a splitting $\mathfrak{g} = \mathfrak{h} \oplus \mathfrak{b}$, where $\mathfrak{h}$ denotes the Lie algebra of some subgroup $H \subset G$ representing the co-adjoint orbit type (H) of points in M. Marle's description is more general than action-group coordinates (which correspond to the case $H = T$), but lacks the same concreteness in the sense that the bundle E^* remains an abstractly defined object.

Dazord and Delzant (1987, Sect. 5) show that in the special 'regular' case $(H = T)$, the bundle E^* trivializes: $E^* \cong G \times U$ $(U \subset \mathbb{R}^k)$. Moreover, they show that U is naturally identifiable with an open subset of a Weyl chamber $\mathfrak{t}_0^*$, the appropriate symplectic structure on $G \times \mathfrak{t}_0^*$ being $-d\Theta_G^*$, where Θ_G^* corresponds, under the identifications discussed in Chap. 2, to the one-form Θ_G defined in 2.3.

The equations of motion 2.10 for action-group coordinates, and the formula of Lemma 2.14 for the Poisson bracket, have not been written out before. We are likewise unaware of explicit constructions of action-group coordinates in concrete examples.

An alternative approach to constructing 'coordinates' in the neighborhood of the orbits of a G action, is to apply the technology of *symplectic cross-sections* (Guillemin et al., 1996; Guillemin and Sternberg, 1984). An application of this technology to action-group coordinates is the main subject of Part 2.

[4] A submanifold is *Lagrangian* if it is simultaneously isotropic and coisotropic, or equivalently if spaces tangent to the submanifold coincide with their symplectic orthogonals.

CHAPTER 4

Naive Averaging

Consider a Hamiltonian G-space P satisfying the hypotheses of Theorem 3.10 and let $H : P \to \mathbb{R}$ be a Hamiltonian of the form

$$H = H_0 + F \ ,$$

where H_0 is G-invariant and F is some perturbation. Then according to the theorem there exists an open set $U \subset \mathfrak{t}_0$ and an equivariant symplectic diffeomorphism $\phi : G \times U \to P$. The equivariance implies that $(H_0 \circ \phi)(g,p) = h(p)$ for some $h : U \to \mathbb{R}$. Thus in action-group coordinates a Hamiltonian system-symmetry breaking problem of the above form translates into the problem of studying a Hamiltonian $H : G \times U \to \mathbb{R}$ of the form

4.1 $$H(g,p) = h(p) + F(g,p) \ .$$

The first task of this section is to describe the dynamics associated with the unperturbed part $H_0(g,p) = h(p)$ of such a Hamiltonian. We will then describe the dynamics associated with an averaged form of the perturbed Hamiltonian H. This will highlight the role played by the symmetry in determining the nature of the perturbed dynamics, as well as motivate the Nekhoroshev estimates developed in subsequent sections.

The unperturbed dynamics

According to 2.10 the equations of motion for a Hamiltonian $H_0(g,p) = h(p)$ are

$$\dot{g} = g\Omega(p) \qquad\qquad \dot{p} = 0 \ ,$$

where $\Omega(p) \equiv \nabla h(p) \in \mathfrak{t}$. We conclude that all integral curves $t \mapsto (g_t, p_t)$ of X_{H_0} are of the form

$$g_t = g_0 e^{t\Omega_0} \qquad\qquad p_t = p_0 \ ,$$

where $\Omega_0 \equiv \Omega(p_0)$. In particular, each G-orbit $G \times \{p_0\}$ ($p_0 \in U$) is an invariant manifold for the flow, and the restriction of the flow to $G \times \{p_0\} \cong G$ is of the form $g_0 \mapsto g_0 e^{t\Omega_0}$. This generalizes the situation in classically integrable systems where one has a foliation by invariant *tori*, each supporting a quasi-periodic motion whose frequency $\Omega(p)$ is delivered by a similar formula.

Corresponding to the identification $G \times \{p_0\} \cong G$ we have an identification of $X_{H_0}|G \times \{p_0\}$ with the left-invariant vector field on G corresponding to $\Omega_0 \in \mathfrak{t} \subset \mathfrak{g}$. This left-invariant vector field is tangent to T and every left coset gT. If we let T act on $G \times U$ according to $q \odot (g,p) \equiv (gq,p)$ ($q \in T$), then these cosets of T in G correspond to T-orbits in the phase space $G \times U$. Thus we have a finer foliation of the phase space $G \times U$ by invariant *tori* (see Fig. 1), but these tori have *strictly less* than half the dimension of phase space in the non-Abelian case.

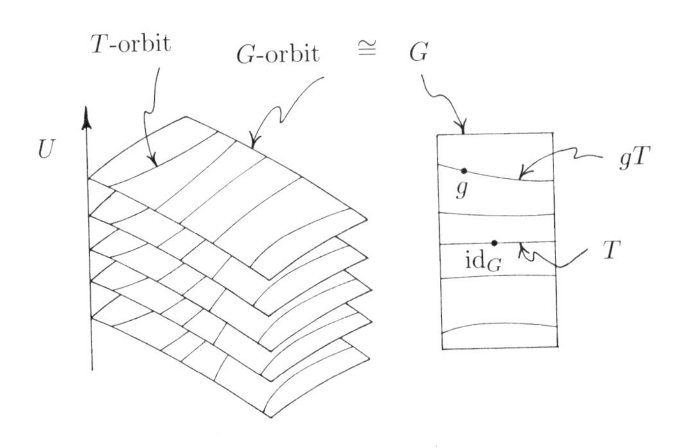

FIGURE 1. Cartoon of the unperturbed motions in $G \times U \subset G \times \mathfrak{t}_0$, depicting the foliation by invariant G-orbits, and the finer foliation by invariant T-orbits.

Let $k = \dim \mathfrak{t}$. The k-dimensional $\mathbb{Z}$-module

$$I_T \equiv \{p \in \mathfrak{t} \mid \exp(p) = \mathrm{id}\} \tag{4.2}$$

is called the *integral lattice* of T in $\mathfrak{t}$. We have:

> *An integral curve of X_{H_0} lying in $G \times \{p_0\}$ is periodic if and only if $\Omega_0 \in \nu I_T \equiv \{\nu \mathbf{n} \mid \mathbf{n} \in I_T\}$ for some $\nu > 0$, in which case a (not necessarily minimal) period is $1/\nu$.*

4.3 Definition. Identify $\mathfrak{t}$ with $\mathbb{R}^k$, by choosing some $\mathbb{Z}$-basis for I_T, and call an element ν of $\mathfrak{t} \cong \mathbb{R}^k$ *irrational* if $\sum_j \nu_j n_j \neq 0$ for all non-zero $n \in \mathbb{Z}^k$.

We leave it to the reader to check that the choice of basis in this definition is immaterial. We have:

> *The integral curves lying on $G \times \{p_0\}$ densely fill each of the T-orbits foliating $G \times \{p_0\}$ if and only if $\Omega_0 \in \mathfrak{t}$ is irrational.*

4.4 Remark. The action of T on $G \times U$ discussed above is in fact *Hamiltonian*, with equivariant momentum map $\mathbf{j}^G : G \times U \to \mathfrak{t}^*$ given by $\mathbf{j}^G(g, p) \equiv (i \circ \varphi)(p)$. See Remark 9.9. The fibers of $\mathbf{j}^G$ are the G-orbits, which constitute the leaves of a coisotropic foliation $\mathcal{F}$ on $G \times U$, by Remark 3.12. Since the space is geometrically integrable, Remark 3.12 also shows that $\mathcal{F}$ is symplectically complete and that the leaves of the associated ω-orthogonal foliation $\mathcal{F}^\omega$ are connected components of fibers of $\mathbf{J}^G$. The fibers of $\mathbf{J}^G$ are in fact connected (indeed they are precisely the orbits of the T action discussed above; see Lemma 4.5.2 below), so that the maps

$$\mathfrak{t}_0^* \xleftarrow{\mathbf{j}^G} G \times U \xrightarrow{\mathbf{J}^G} \mathfrak{g}^*$$

constitute a concrete realization of the foliations $\mathcal{F}$ and $\mathcal{F}^\omega$ as fibers of a dual pair[1].

[1] A pair of Poisson maps $Q_1 \xleftarrow{\rho_1} P \xrightarrow{\rho_2} Q_2$ (Q_1, Q_2 Poisson manifolds) is a *dual pair* if the tangent maps $T\rho_1$ and $T\rho_2$ have ω-orthogonal kernels.

Since $\mathcal{F}$ is a symplectically complete coisotropic foliation with compact fibers, one may construct (see, e.g., Dazord and Delzant (1987)) an atlas of partial action-angle coordinate charts on $G \times U$. In each chart

$$U_1 \xrightarrow{\sim} \mathbb{T}^k \times U_2 \qquad (U_1 \subset G \times U, \quad U_2 \subset \mathbb{R}^k \times \mathbb{R}^{2(n-k)})$$

the foliation by T-orbits will be represented by the trivial foliation

$$\left(\mathbb{T}^k \times \{y\} \right)_{y \in U_2} \ .$$

But if G is a non-Abelian compact connected Lie group, the bundle $G \to G/T$ is topologically *nontrivial*. Since the T-orbits lying in a G-orbit $G \times \{p\} \cong G$ $(p \in U)$ correspond to left cosets of T in G, we conclude that no partial action-angle coordinate chart can be constructed in the non-Abelian case *such that it contains an entire neighborhood of some G-orbit.*

The averaged dynamics

Suppose that a Hamiltonian system with Hamiltonian H_0 admits a foliation $\mathcal{F}$ by closed (compact and boundaryless) invariant manifolds, and assume that the restricted flow on a generic leaf is ergodic[2]. The familiar example is the foliation by invariant tori of an integrable system in action-angle coordinates, provided the Hamiltonian $H_0(q,p) = h(p)$ is *nondegenerate*. Furthermore, assume (as in this example) that each leaf of $\mathcal{F}$ supports a probability measure invariant with respect to the restricted Hamiltonian flow, and varying from leaf to leaf in some smooth fashion. Given a perturbed Hamiltonian H, one can form a 'first order approximation' $\bar{H}$ to H, constructed by averaging H over the leaves of $\mathcal{F}$ using the measures. One then expects the flow of $\bar{H}$ to be a reasonable approximation to that of H (at least better than that of H_0). Turning this heuristic expectation into rigorous theorems is, more or less, the principal preoccupation of the classical perturbation theory[3]. The necessity of ergodicity is well-known.

We now carry out the averaging procedure just outlined in the case of an integrable Hamiltonian system in action-group coordinates. Note that in the example of action-angle coordinates mentioned above the averaged Hamiltonian $\bar{H}$ is itself *integrable*, and therefore not terribly interesting.

Assume that the unperturbed Hamiltonian $H_0(g,p) = h(p)$ in 4.1 is nondegenerate, i.e. that the map $p \mapsto \Omega(p) : U \to \mathfrak{t}$ is a local diffeomorphism. Then for almost any $p \in U$ (in the sense of Lebesgue measure), the frequency $\Omega(p)$ is irrational (in the sense of 4.3), in which case the Hamiltonian flow on the G-orbit $G \times \{p\}$ is ergodic on each torus $gT \times \{p\}$ $(g \in G)$. We therefore choose the leaves of $\mathcal{F}$ in our preceeding discussion to be the orbits of the T-action $q \odot (g,p) \equiv (gq,p)$ discussed above, and define

$$\bar{H}(g,p) \equiv \int_T H(gq,p)\,d\mu(q) \ .$$

Here μ denotes the unique translation-invariant probability measure on the torus T. By construction, $\bar{H}$ is T-invariant. Our objective is to Poisson-reduce (see, e.g., Marsden (1992)) the dynamics of $\bar{H}$ to dynamics on an appropriate realization of the quotient space $(G \times U)/T$.

---

[2]By 'generic', we mean with respect to some appropriate Borel measure on the factor space, which we suppose is Hausdorff.

[3]See also the introduction to the book by Lochak and Meunier (1988).

Recall that the action of G on $G \times \mathfrak{t}_0$ admits an equivariant momentum map $\mathbf{J}^G : G \times \mathfrak{t}_0 \to \mathfrak{g}^*$ defined by 3.3. With the assistance of Theorem 1.2.6 and Theorem 1.2.7, it is not difficult to prove the following:

4.5 Lemma.

1. *The fibers of $\mathbf{J}^G$ are the T-orbits.*
2. *If $\mathcal{O} \subset \mathfrak{g}^*_{\mathrm{reg}}$ is a co-adjoint orbit, then $(\mathbf{J}^G)^{-1}(\mathcal{O})$ is some G-orbit $G \times \{p\}$ $(p \in \mathfrak{t}_0)$.*

Lemma 4.5.1 says that we have a natural way of identifying the abstract quotient $(G \times U)/T$ with $V \equiv \mathbf{J}^G(G \times U) \subset \mathfrak{g}^*$, the momentum map $\mathbf{J}^G : G \times U \to V$ being a realization of the natural projection $G \times U \to (G \times U)/T$.

Being an *equivariant* momentum map, $\mathbf{J}^G$ is also a *Poisson* map, if we equip $\mathfrak{g}^*$ with the positive Lie-Poisson structure $\{\cdot,\cdot\}_+$ (see, e.g., Marsden and Ratiu (1994, Chapter 10)). If u is a smooth function on $\mathfrak{g}^*$, then by definition its corresponding *Hamiltonian vector field* X_u on $\mathfrak{g}^*$ is the vector field satisfying the equation $X_u \lrcorner df = \{f, u\}_+$ for all smooth $f : \mathfrak{g}^* \to \mathbb{R}$. Hamiltonian vector fields on a Poisson manifold are always tangent to the symplectic leaves, which in this case are the co-adjoint orbits. If a function is constant on a leaf, its Hamiltonian vector field vanishes.

The functions $\bar{H}$, H_0 and $\bar{F}$ on $G \times U$ are T-invariant. They therefore drop, via $\mathbf{J}^G : G \times U \to V$, to functions $\bar{H}'$, H_0' and $\bar{F}'$ on $V \subset \mathfrak{g}^*$. Since H_0 is G-invariant, 4.5.2 guarantees that H_0' is constant on the co-adjoint orbits, and hence that $X_{H_0'} = 0$. Therefore,

$$X_{\bar{H}'} = X_{\bar{F}'} \ .$$

Since $\mathbf{J}^G$ is a Poisson map, the vector fields $X_{\bar{H}}$ and $X_{\bar{H}'}$ are $\mathbf{J}^G$-related. Therefore, $\mathbf{J}^G$ *maps integral curves of $X_{\bar{H}}$ onto integral curves of $X_{\bar{F}'}$.*

If $F = 0$, then $X_{\bar{F}'} = 0$. This is consistent with Noether's theorem, which states that $\mathbf{J}^G(x_t) \in \mathfrak{g}^*$ is constant for all integral curves $t \mapsto x_t$ of $X_{\bar{H}}$, if $\bar{H}$ is G-invariant.

If $F \neq 0$, then $X_{\bar{F}'}$ need not be trivial, or even integrable, but must nevertheless remain tangent to the co-adjoint orbits. It follows from 4.5.2 that the G-orbits $G \times \{p\}$ $(p \in U)$ persist as invariant manifolds of $X_{\bar{H}}$. In other words, *the action variables $p \in U$ are integrals of motion for the averaged Hamiltonian $\bar{H}$.*

Although one cannot show that the action variables persist as constants of the motion in the fully perturbed (unaveraged) system, one can derive, under certain conditions, exponential bounds on their evolution of Nekhoroshev type. We turn to such estimates in the following chapters.

On the other hand, the non-trivial dynamics on the co-adjoint orbits created by an $F \neq 0$ means that a solution curve $t \mapsto x_t$ of $X_{\bar{H}}$, which must lie on some G-orbit $G \times \{p\} \cong G$, moves from one T-orbit to another, in a fashion determined by $\bar{F}'$, and at a speed of order $1/\epsilon$, if ϵ is the size of the perturbation F. These 'fast' motions correspond to those pointed out by Fassò (1995) in the context of symplectically complete isotropic foliations, which we described in the Introduction. In particular (see 4.4), one cannot guarantee that a trajectory of the averaged Hamiltonian will stay in a partial action-angle coordinate chart longer than a time of order $1/\epsilon$. This is far shorter than the exponential time scales one hopes to establish for the action variables (in the fully perturbed system), explaining the shortcoming of partial action-angle coordinate charts described in the Introduction.

4.6 Remark. We now elaborate on Remark 2.11's claim that $G \times \mathfrak{g}$ is poorly suited as a system of coordinates for studying perturbations to left-invariant Hamiltonians on T^*G (G non-Abelian, compact and connected). We assume the reader has read our comments under 'The average dynamics' above. In the coordinates $(g, \mu) \in G \times \mathfrak{g}$ the unperturbed equations of motion are (see 2.11)

$$\dot{g} = g\nabla h(\mu) \ , \qquad \dot{\mu} = [\mu, \nabla h(\mu)] \ .$$

At first, the existence of a dynamically invariant foliation is not obvious but, writing $m \equiv \mathrm{Ad}_g \, \mu = g\mu g^{-1}$, we obtain

$$\dot{g} = g\nabla h(g^{-1}mg) \ , \qquad \dot{m} = 0 \ .$$

That is, in the new coordinates $(g, m) \in G \times \mathfrak{g}$, there is a foliation $\mathcal{F}$ by invariant manifolds of the form $G \times \{m\}$ $(m \in \mathfrak{g})$. Unfortunately, the dynamics restricted to a leaf $G \times \{m\}$, which is governed by the equation $\dot{g} = g\nabla h(g^{-1}mg)$, need not be integrable, unless $\dim G = 3$ (if $\dim G < 3$, then G is Abelian). Even if $\dim G = 3$ the restricted dynamics is generally not ergodic. Of course, there may be a finer foliation by invariant manifolds that generically support ergodic flows (the Euler-Poinsot rigid body is a case in point). However, *the geometry of this foliation depends on the particular Hamiltonian h.* Since the coordinates (g, m) (or (g, μ)) do not see this foliation, they will not function well in a perturbation analysis.

In fact the integrability obtained in the $\dim G = 3$ case amounts to mere *dynamic* integrability (see 3.7). An alternative strategy for dealing with such systems is to convert to *geometric* integrability (by exploiting knowledge about the particular Hamiltonian h) and then to construct action-group coordinates. See Remark 3.7.

CHAPTER 5

An Abstract Formulation of Nekhoroshev's Theorem

When $G = \mathbb{T}^n$, the action-group model space $G \times \mathfrak{t}_0$ is just ordinary action-angle coordinates $\mathbb{T}^n \times \mathbb{R}^n$. In this case Nekhoroshev's theorem (Nekhoroshev, 1977) asserts that for all sufficiently small perturbations F in the Hamiltonian 4.1, the action variables p evolve, in some sense, exponentially slowly. This is provided that the Hamiltonian is real-analytic and satisfies a certain 'convexity' assumption. See, e.g., Lochak (1992) for a precise statement.

Our objective is to generalize Nekhoroshev's result to the case in which G is a general compact connected Lie group. We refer to this as a problem in *non-canonical* perturbation theory, since the symplectic structure ω_G on $G \times \mathfrak{t}_0$ is non-canonical when G is non-Abelian (in the sense given in the Introduction).

Many of the techniques involved in generalizing Nekhoroshev's theorem are not new. For example, one can still make the use of Lie transforms to effect symplectic coordinate changes. Indeed, once one has certain 'basic estimates' (e.g., bounds on Poisson brackets), the proof of Nekhoroshev's theorem, although non-trivial, requires only basic properties of the underlying coordinate system and its symplectic structure. In this chapter we formalize these basic requirements (Assumptions A–C below) and formulate a corresponding 'abstract' version of Nekhoroshev's theorem (Theorem 5.8). As the proof is, for the most part, a rewriting of Lochak's (1992) proof of Nekhoroshev's theorem for conventional action-angle coordinates (see also (Lochak, 1993)), it is relegated to Appendix A. In Chap. 6 we will demonstrate that our abstraction applies to action-group coordinates.

Throughout this chapter, the reader may find it helpful to keep the example of conventional action-angle coordinates in mind. In what follows this corresponds to the choice $G = \mathbb{T}^n$ and $\mathfrak{t} = \mathbb{R}^n$.

Notation

Let M be a compact C^0-manifold with (possibly void) boundary whose interior $\operatorname{int} M$ is a complex (resp. real-analytic) manifold, and let V be a complex (resp. real-analytic) vector space. Then we denote by $\mathcal{A}^V(M)$ the space of continuous maps $u : M \to V$ that are holomorphic (resp. real-analytic) on $\operatorname{int} M$. If $|\cdot|$ is a norm on V, then the space $\mathcal{A}^V(M)$ is equipped with the norm

$$\|u\| \equiv \sup_{x \in M} |u(x)| \ .$$

Preliminary notions

An essential ingredient in any proof of Nekhoroshev's theorem is the complexification of the coordinates, and the analytic continuation of the Hamiltonian (which must be assumed real-analytic). The following definition formalizes in a precise way the notion that we require.

5.1 Definition. Let G be a real-analytic manifold (without boundary), and suppose that there exists a nested family $(G^\sigma)_{0\leqslant\sigma\leqslant\bar\sigma<\infty}$ of sets ($\sigma < \sigma' \Rightarrow G^\sigma \subset G^{\sigma'}$) with $G^0 = G$ and satisfying the conditions:

1. G^σ is a C^0-manifold with boundary for all $\sigma \neq 0$.
2. $\operatorname{int} G^\sigma$ is a complex manifold for all $\sigma \neq 0$, the inclusion $\operatorname{int} G^\sigma \hookrightarrow \operatorname{int} G^{\bar\sigma}$ being holomorphic for all $\sigma \neq 0$.
3. Any real-analytic tensor on $G^0 = G$ has, for some $\sigma > 0$, a holomorphic extension to $\operatorname{int} G^\sigma$ and a continuous extension to G^σ.

Then we call the family $(G^\sigma)_{0\leqslant\sigma\leqslant\bar\sigma}$ a *local complexification* of G.

5.2 Remark. Local complexifications exist under rather general conditions, and are unique up to an appropriate equivalence (Whitney and Bruhat, 1959).

See 5.5 below for an example of a local complexification in the above sense. (In Chap. 6 we shall write down a local complexification explicitly in the case that G is an arbitrary compact connected Lie group.)

5.3 Definition. Let V be a real (resp. complex) normed vector space and let $U \subset V$ be open. Let a real-analytic (resp. holomorphic) map $h : U \to \mathbb{R}$ (resp. $\mathbb{C}$) be called (m, M)-*convex* if there exists $m, M > 0$ such that

$$\left.\begin{array}{rcl} D^2h(p)(v,v) & \geqslant & m|v|^2 \\ |D^2h(p)(u,v)| & \leqslant & M|u||v| \end{array}\right\} \qquad \forall u,v\in V \ , \ \forall p \in U \ .$$

The abstract set-up

Let G be a compact real-analytic manifold (not necessarily a Lie group) and $(G^\sigma)_{0\leqslant\sigma\leqslant\bar\sigma}$ a local complexification, with G^σ compact for all σ. Let $\mathfrak{t}$ be a finite dimensional real vector space with inner product $(a,b)\mapsto a\cdot b$, and corresponding norm $|\cdot|$.

We assume that for some open set $\mathfrak{t}_0 \subset \mathfrak{t}$ the real-analytic manifold $G \times \mathfrak{t}_0$ is equipped with a symplectic two-form ω.

For $p \in \mathfrak{t}_0$ write

$$B_R(p) \equiv \{p' \in \mathfrak{t} \mid |p' - p| \leqslant R\} \ .$$

Extend the inner product on $\mathfrak{t}$ to a $\mathbb{C}$-bilinear form on $\mathfrak{t}^{\mathbb{C}} \equiv \mathfrak{t} \otimes_{\mathbb{R}} \mathbb{C}$. A norm on $\mathfrak{t}^{\mathbb{C}}$ is then given by $|a\otimes 1 + b \otimes i| \equiv \sqrt{a\cdot a + b\cdot b}$. Write

$$B_R^\rho(p) \equiv \{p' \in \mathfrak{t}^{\mathbb{C}} \mid |p' - B_R(p)| \leqslant \rho\} \ .$$

Fix $\bar p \in \mathfrak{t}_0$ and positive constants $\bar\sigma$ and $\bar\rho$, and write $B \equiv B_{\bar\rho}(\bar p)$ and $B^{\bar\rho} \equiv B_{\bar\rho}^{\bar\rho}(\bar p)$, so that $B^{\bar\rho}$ is a complex neighborhood of the real ball B. Assume that $\bar\sigma$ and $\bar\rho$ are small enough that ω has a holomorphic extension to a (complex) bounded symplectic form on $G^{\bar\sigma}\times B^{\bar\rho}$. By *bounded* we mean that the induced Poisson bracket satisfies the following property: for all u and v in $\mathcal{A}^{\mathbb{C}}(G^{\bar\sigma}\times B^{\bar\rho})$, the bracket $\{u,v\}$ is assumed bounded on $\operatorname{int}(G^{\bar\sigma}\times B^{\bar\rho})$, with a continuous extension to $G^{\bar\sigma}\times B^{\bar\rho}$, i.e.

$$u, v \in \mathcal{A}^{\mathbb{C}}(G^{\bar\sigma}\times B^{\bar\rho}) \Rightarrow \{u,v\} \in \mathcal{A}^{\mathbb{C}}(G^{\bar\sigma}\times B^{\bar\rho}) \ .$$

Consider a Hamiltonian of the form

$$H(g,p) = h(p) + F(g,p) \ , \tag{5.4}$$

where $h \in \mathcal{A}^{\mathbb{R}}(B)$ and $F \in \mathcal{A}^{\mathbb{R}}(G \times B)$. Assume that h is (m', M')-convex on int B for some positive constants m' and M', and that $\bar{\sigma}$ and $\bar{\rho}$ are small enough that:

1. h and F have holomorphic extensions $h \in \mathcal{A}^{\mathbb{C}}(B^{\bar{\rho}})$, $F \in \mathcal{A}^{\mathbb{C}}(G^{\bar{\sigma}} \times B^{\bar{\rho}})$.
2. There exists $m, M > 0$ such that h is (m, M)-convex on int $B^{\bar{\rho}}$.

Fundamental assumptions

Make the following crucial assumption about the nature of the Hamiltonian vector fields defined by ω:

Assumption A (Existence of a 'period lattice'). Let X_u denote the (complex) Hamiltonian vector field corresponding to a function $u \in \mathcal{A}^{\mathbb{C}}(G^{\bar{\sigma}} \times B^{\bar{\rho}})$. Assume that there exists a lattice $\mathfrak{t}^{\mathbb{Z}} \subset \mathfrak{t}$, whose dimension (over $\mathbb{Z}$) equals the dimension of $\mathfrak{t}$ (over $\mathbb{R}$), with the following property. Suppose $W \in \mathcal{A}^{\mathbb{C}}(G^{\bar{\sigma}} \times B^{\bar{\rho}})$ is of the form $W(g,p) = w(p)$. Then if $\nabla w(p) \in \nu\mathfrak{t}^{\mathbb{Z}} \equiv \{\nu\mathbf{n} \mid \mathbf{n} \in \mathfrak{t}^{\mathbb{Z}}\}$ for some $p \in B^{\bar{\rho}}$ and $\nu > 0$, we require that all integral curves of X_W beginning in $G^{\sigma} \times \{p\}$ ($\sigma \in [0, \bar{\sigma}]$ arbitrary) remain in $G^{\sigma} \times \{p\}$ for all time and are **periodic** with (not necessarily minimal) period $1/\nu$. (If $\nabla w(p) = 0$, this condition implies that $G^{\bar{\sigma}} \times \{p\}$ is a manifold of equilibria for X_W.)

5.5 Example (action-angle coordinates). Take $G = \mathbb{T}^n \equiv \mathbb{R}^n/2\pi\mathbb{Z}^n$, $\mathfrak{t} = \mathfrak{t}_0 = \mathbb{R}^n$ and $\omega = \Sigma dq_j \wedge dp_j$. Identifying $\mathbb{T}^n$ with $\{(z_1, \dots, z_n) \in \mathbb{C}^n \mid |z_j| = 1\ \forall j\}$, we see that a local complexification $(G^{\sigma})_{0 \leqslant \sigma \leqslant \bar{\sigma}}$ of G is given by $G^{\sigma} \equiv \{(z_1, \dots, z_n) \in \mathbb{C}^n \mid |z_j| \in [1-\sigma, 1+\sigma]\ \forall j\}$, $\bar{\sigma} \equiv \frac{1}{2}$. We satisfy Assumption A if we take $\mathfrak{t}^{\mathbb{Z}} = 2\pi\mathbb{Z}^n$.

Let $k \equiv \dim \mathfrak{t}$ and fix a basis $\{\beta_1, \dots, \beta_k\}$ for $\mathfrak{t}^{\mathbb{Z}}$ (which we do *not* require to be orthogonal with respect to the inner product on $\mathfrak{t}$). Let $c_1 > 0$ be a constant (depending only on $\mathfrak{t}^{\mathbb{Z}}$, the inner product on $\mathfrak{t}$, and the choice of basis) such that

5.6
$$|\tau_1\beta_1 + \dots + \tau_k\beta_k| \leqslant c_1\sqrt{k} \max_{0 \leqslant j \leqslant k} |\tau_j| \qquad (\tau_j \in \mathbb{R}) \ .$$

It is clear that such a constant exists. (It will be estimated explicitly in the case of action-group coordinates in Chap. 6.)

Let $p_* \in \operatorname{int} B$ and $0 \leqslant r \leqslant 1$ be free parameters and define the following family of (complex) neighborhoods of $G \times \{p_*\}$ (see Fig. 1):

$$D_\gamma(p_*, r) \equiv G^{\gamma\bar{\sigma}} \times B_{r\bar{\rho}}^{\gamma r\bar{\rho}}(p_*) \qquad (0 \leqslant \gamma \leqslant 1) \ .$$

So that these domains are well-defined subsets of $G^{\bar{\sigma}} \times B^{\bar{\rho}}$, the parameters p_* and r must be subject to the condition

5.7
$$r \leqslant 1 - \frac{|p_* - \bar{p}|}{\bar{\rho}} \ .$$

Note that $D_0(p_*, r) = G \times B_{r\bar{\rho}}(p_*) \subset G \times B$ is a *real* neighborhood of $G \times \{p_*\} = D_0(p_*, 0)$, and that $D_1(\bar{p}, 1) = G^{\bar{\sigma}} \times B^{\bar{\rho}}$ is the complex domain on which the Hamiltonian H is defined by hypothesis. We denote the supremum norm on $\mathcal{A}^{\mathbb{C}}(D_\gamma(p_*, r))$ by $\|\cdot\|_\gamma^{p_*, r}$.

The remaining hypotheses we require are:

Assumption B (On 'times of validity'). There exists $c_2 > 0$, independent of m, M, γ, p_* and r, such that for all $u \in \mathcal{A}^{\mathbb{C}}(D_\gamma(p_*, r))$, $0 < \delta \leqslant \gamma$ and $|t| \leqslant t_0 \equiv$

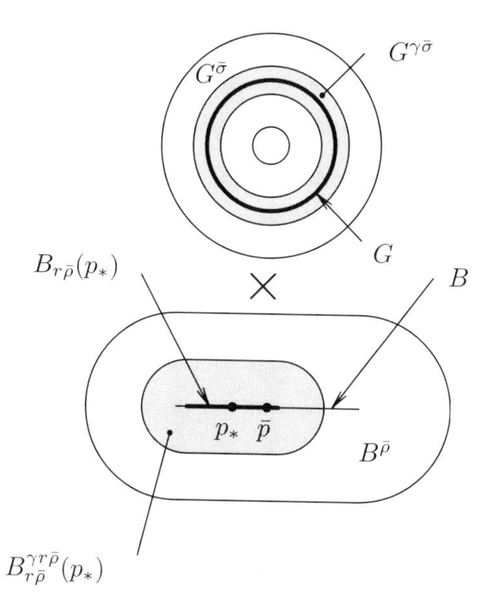

FIGURE 1. Schematic of the domain $D_\gamma(p_*, r) \equiv G^{\gamma\bar{\sigma}} \times B^{\gamma r\bar{\rho}}_{r\bar{\rho}}(p_*)$ (shaded).

$c_2\bar{\sigma}\bar{\rho}\delta^2 r/\|u\|^{p_*,r}_\gamma$, every integral curve $\tau \mapsto (g_\tau, p_\tau)$ of X_u beginning at a point $(g_0, p_0) \in D_{\gamma-\delta}(p_*, r)$ is well-defined and satisfies

$$(g_t, p_t) \in D_{\gamma-\delta/2}(p_*, r) \text{ with } |p_t - p_0| \leqslant \left(\frac{|t|}{t_0}\right)\left(\frac{\delta}{2}\right) r\bar{\rho} \ .$$

Assumption C (The existence of a 'Cauchy inequality' for Poisson brackets). There exist constants $c_3 > 0$ and $c_4 > 0$, independent of m, M, γ, p_* and r, such that for all $0 < \delta \leqslant \gamma$ and all $u, v, W \in \mathcal{A}^{\mathbb{C}}((D_\gamma(p_*, r))$ with W of the form $W(g, p) = w(p)$ one has the estimates

$$\|\{u, v\}\|^{p_*,r}_{\gamma-\delta} \leqslant \left(\frac{c_3}{\bar{\sigma}\bar{\rho}}\right) \frac{\|u\|^{p_*,r}_\gamma \|v\|^{p_*,r}_\gamma}{\delta^2 r}$$

$$\|\{W, v\}\|^{p_*,r}_{\gamma-\delta} \leqslant c_4 \left(\sup_{p \in B^{\gamma r\bar{\rho}}_{r\bar{\rho}}(p_*)} \left|\nabla w(p)\right| \right) \frac{\|v\|^{p_*,r}_\gamma}{\delta^2} \ .$$

(Recall that $\{u, v\} \equiv X_v \lrcorner X_u \lrcorner \omega$.)

Nekhoroshev's theorem

Write $\Omega(p) = \nabla h(p) \in \mathfrak{t}$ and define the 'time scales'

$$T_m \equiv \frac{1}{\bar{\rho} m} \qquad T_M \equiv \frac{1}{\bar{\rho} M}$$

$$T_\Omega \equiv \left(\sup_{p \in B^{\bar{\rho}}} |\Omega(p)|\right)^{-1} \qquad T_{h'''} \equiv \left(\bar{\rho}^2 \sup_{p \in B} \sup_{u,v,w \in \mathfrak{t}} |D^3 h(p)(u, v, w)|\right)^{-1}$$

and the associated 'dimensionless constants'

$$c_5 \equiv \frac{T_{h'''}}{T_m} \qquad c_6 \equiv \frac{T_M}{T_\Omega} \qquad c_7 \equiv \frac{T_M}{T_m} = \frac{m}{M} \ .$$

We allow $T_{h'''} = \infty$ (in which case $c_5 = \infty$). A constant with the physical dimensions of the Hamiltonian H_0 is

$$E \equiv \frac{\bar{\sigma}\bar{\rho}}{T_M} \ .$$

Recall that k denotes the dimension of the real vector space $\mathfrak{t}$.

5.8 Theorem (Nekhoroshev-Lochak (abstract form)).
Consider the scenario desribed in this section with Hamiltonian H of the form 5.4, under the Assumptions A, B and C above. Let $\|\cdot\|$ denote the supremum norm on $\mathcal{A}^{\mathbb{C}}(G^{\bar{\sigma}} \times B^{\bar{\rho}})$ and define the 'perturbation parameter'

$$\epsilon \equiv \frac{\|F\|}{E} \ .$$

Then there exist positive constants $a = a(k)$, $b = b(k)$, c, $\epsilon_0 = \epsilon_0(k, c_1, \ldots, c_7)$, $t_0 = t_0(c_1, \ldots, c_7)$ and $r_0 = r_0(c_1, \ldots, c_7)$, such that for all $\epsilon \leqslant \epsilon_0$ every (real) solution curve $t \mapsto (g_t, p_t) \in G \times B$ of X_H with $p_0 = \bar{p}$ obeys the exponential stability estimate

$$\frac{|t|}{T_\Omega} \leqslant t_0 \exp(c\epsilon^{-a}) \ \Rightarrow \ \frac{|p_t - p_0|}{\bar{\rho}} \leqslant r_0 \epsilon^b \ .$$

A proof of 5.8, based largely on Lochak (1992) and Lochak (1993), is outlined in Appendix A. Estimates for the constants $a, b, c, \epsilon_0, t_0, r_0$ obtained in Appendix A are summarized below. For pedagogical reasons, we have not attempted to optimize the exponents a and b (these *are* optimized in Lochak and Neishtadt (1992) and Lochak (1993)). Exponent values $a = \frac{1}{2k}$ and $b = \frac{1}{2k}$ should be possible (c.f. $a = \frac{1}{2n}$ and $b = \frac{1}{2n}$ in opi cit., where n is the number of degrees of freedom). For more information see Appendix A.

5.9 Addendum. *In Theorem 5.8 above one may take*

$$a = \frac{1}{4(1+k)} \ , \qquad b = \frac{1}{2(1+k)} \ , \qquad c = 1$$

$$\epsilon_0 = \min\left\{ \left(\frac{8}{{c_7'}^3 l_1 l_2} \right)^{1/b}, \left(\frac{1}{2{c_7'}^2 l_1 l_2} \right)^{1/b} \right\} \ ,$$

$$t_0 = \frac{3{c_7'}^2}{8ec_4} \ , \qquad r_0 = 2\left(1 + \frac{1}{c_7'}\right) \frac{{c_7'}^3 l_1 l_2}{16} \ ,$$

where

$$l_4 \equiv \frac{{c_7'}^2 l_2}{4} \left(\frac{c_7 {c_7'}^3 l_2}{16 c_1 \sqrt{k}} \right)^k \ , \qquad l_3 \equiv \min\left\{ c_7 c_5', 2c_1\sqrt{k} \right\} \ ,$$

$$l_2 \equiv \min\left\{ c_2, \frac{e}{2c_4 + 5c_3} \right\} \ , \qquad l_1 \equiv \max\left\{ 1, \frac{192\bar{\sigma}}{c_7}, \left(\frac{1}{l_4} \right)^{k+1/2} \right\} \ ,$$

$$c_7' \equiv \frac{\sqrt{5} - 2}{2} c_7 \ , \qquad c_5' \equiv \min\{\frac{c_5}{2}, 1\} \ .$$

CHAPTER 6

Applying the Abstract Nekhoroshev Theorem to Action-Group Coordinates

6.1 Proposition. *The abstract formulation 5.8 of Nekhoroshev's theorem applies in the case that G is a compact connected Lie group and where $\mathfrak{t}$, $\mathfrak{t}_0$ and $\omega = \omega_G$ have the meanings given in Chap. 2.*

The objective of this chapter is to prove the above result. Recall that up to a finite covering any compact connected Lie group is the direct sum of a semisimple and Abelian (toral) factor (see, e.g., Bröcker and tom Dieck (1985, V8.1)). Although 6.1 holds for any compact connected Lie group G, we shall only prove it in the case that G is semisimple. The Abelian case is already covered by Nekhoroshev's original statement. Furthermore, it turns out that there is essentially no 'coupling' between the Abelian and semisimple factors in the general case. To simultaneously keep track of both an Abelian and semisimple factor just introduces uninstructive bookkeeping to the calculations.

For the remainder of this section G denotes a compact connected semisimple Lie group.

In this section we shall make use of basic properties of semisimple Lie algebras, as well as their *root* and *inverse root* systems. We refer the reader to Bröcker and tom Dieck (1985) for a discussion of the relationship between these roots and the Weyl chambers, and for further background. Here a root will always mean a *real* root, in the sense of, e.g., V1.3, op. cit.

The local complexification of G

It is now convenient to fix the choice of Ad-invariant inner product on $\mathfrak{g}$ introduced in Chap. 1, which until now has been completely arbitrary. Define

$$\xi \cdot \eta \equiv - \operatorname{Trace}(\operatorname{ad}_\xi \circ \operatorname{ad}_\eta) \qquad (\xi, \eta \in \mathfrak{g}) \ .$$

Minus one times this product is known as the *Killing form* of $\mathfrak{g}$. With G as above, the Killing form is symmetric, Ad-invariant and negative definite. Remember that we have been denoting the unique $\mathbb{C}$-bilinear extension of our product to $\mathfrak{g}^{\mathbb{C}}$ by $\xi \cdot \eta$ also. We continue to denote the norm on $\mathfrak{g}$ induced by the inner product (and its obvious extension to $\mathfrak{g}^{\mathbb{C}}$) by $|\cdot|$.

Recall that we assume that G is realized as a subgroup of $\mathrm{SO}(n_G, \mathbb{R})$, so that we may view $\mathfrak{g}^{\mathbb{C}}$ as a subalgebra of $\mathbb{C}^{n_G \times n_G}$. The latter Lie algebra is equipped with a natural symmetric $\mathbb{C}$-bilinear form, namely the *Schur-Hadamard product*

$$S(A, B) \equiv \frac{1}{n_G} \operatorname{Trace}(AB^T) \qquad (A, B \in \mathbb{C}^{n_G \times n_G}) \ .$$

An associated norm on $\mathbb{C}^{n_G \times n_G}$ is given by

$$|A|_S \equiv \sqrt{S(A,\bar{A})} = \sqrt{\frac{1}{n_G}\,\mathrm{Trace}(A\bar{A}^T)}\ ,$$

where a bar denotes complex conjugation. Note that as $G \subset \mathrm{SO}(n_G,\mathbb{R})$,

6.2 $$|g|_S = 1 \qquad (g \in G)$$

6.3 and $$|gA|_S = |A|_S = |Ag|_S \qquad (g \in G, A \in \mathbb{C}^{n_G \times n_G})\ .$$

By writing things out with respect to the standard basis of $\mathbb{C}^{n_G}$ and applying the Schwartz inequality several times, one can derive the estimate

6.4 $$|AB|_S \leqslant \sqrt{n_G}|A|_S|B|_S \qquad (A,B \in \mathbb{C}^{n_G \times n_G})\ .$$

If $\bar{\sigma} > 0$ is chosen sufficiently small, then a well-defined local complexification $(G^\sigma)_{0 \leqslant \sigma \leqslant \bar{\sigma}}$ of G is given by

6.5 $$G^\sigma \equiv \{g \in G^{\mathbb{C}} \mid |g - G|_S \leqslant \sigma\}\ .$$

It will be convenient to assume that $\bar{\sigma} \leqslant 1$. Also, with $\bar{p} \in \mathfrak{t}_0$ fixed, we assume $\bar{\rho} > 0$ is a constant small enough that

6.6 $$B_{2\bar{\rho}}(\bar{p}) \subset \mathfrak{t}_0\ .$$

We need to establish the relation between $|\cdot|$ (the norm on $\mathfrak{g}^{\mathbb{C}}$ determined by the Killing form) and the restriction to $\mathfrak{g}^{\mathbb{C}}$ of $|\cdot|_S$. Using the fact that $G \subset \mathrm{SO}(n_G,\mathbb{R})$, it is not difficult to show that $S(\cdot,\cdot)$ is invariant with respect to the adjoint action of G on $\mathbb{C}^{n_G \times n_G}$. In particular its restriction to $\mathfrak{g}$ is Ad-invariant. But as G is semisimple, there is up to a constant only one Ad-invariant inner product on $\mathfrak{g}$, namely the Killing form (see op. cit.). By the uniqueness of extensions of $\mathbb{R}$-bilinear forms on $\mathfrak{g}$ to $\mathbb{C}$-bilinear forms on $\mathfrak{g}^{\mathbb{C}}$, it follows that

6.7 $$S(\xi,\eta) = k_1 \xi \cdot \eta \qquad (\xi,\eta \in \mathfrak{g}^{\mathbb{C}})\ ,$$

for some $k_1 > 0$.

The constant k_1 depends on the realization of $G \subset \mathrm{SO}(n_G,\mathbb{R})$ and is the only constant we leave implicitly defined. It is easily computed on a case-by-case basis.

From 6.7 it follows that

6.8 $$|\xi|_S = \sqrt{k_1}|\xi| \qquad (\xi \in \mathfrak{g}^{\mathbb{C}})\ .$$

In particular we deduce from 6.4 and 6.8 the estimate

6.9 $$|g\xi|_S \leqslant k_2|g|_S|\xi| \qquad (g \in G^{\mathbb{C}}, \xi \in \mathfrak{g}^{\mathbb{C}})\ ,$$

where

6.10 $$k_2 \equiv \sqrt{n_G k_1}\ .$$

Existence of a period lattice (Assumption A)

Define

$$\bar{k} \equiv (\dim G - \mathrm{rank}\, G)/2 \geqslant k \equiv \dim \mathfrak{t} = \mathrm{rank}\, G\ .$$

Then with respect to the fixed chamber $\mathfrak{t}_0$, $\mathfrak{g}$ has $\bar{k}$ positive real roots

$$\alpha_1, \dots, \alpha_k, \alpha_{k+1}, \dots, \alpha_{\bar{k}} \in \mathfrak{t}^*\ .$$

We assume that these have been ordered so that $\{\alpha_1, \dots, \alpha_k\}$ is the basis of $\mathfrak{t}^*$ consisting of the indecomposable elements of $\{\alpha_1, \dots, \alpha_{\bar{k}}\}$. The corresponding inverse roots will be denoted $\alpha_1^*, \dots, \alpha_{\bar{k}}^* \in \mathfrak{t}$. They are given by

6.11
$$\alpha_j^* = \frac{2\kappa^{-1}(\alpha_j)}{B(\alpha_j, \alpha_j)} \qquad (1 \leqslant j \leqslant \bar{k}) \ ,$$

where $\kappa : \mathfrak{t} \to \mathfrak{t}^*$ is the isomorphism $\langle \kappa(a), b \rangle \equiv a \cdot b$, and $B(\,\cdot\,,\,\cdot\,)$ is the induced inner product on $\mathfrak{t}^*$. *It is a fact that the inverse roots lie on the integral lattice I_T of T in* $\mathfrak{t}$ (defined by 4.2).

Suppose $W \in \mathcal{A}^{\mathbb{C}}(G^{\bar{\sigma}} \times B^{\bar{\rho}})$ is of the form $W(g,p) = w(p)$. Then, by the equations of motion 2.10, an integral curve $t \mapsto (g_t, p_t)$ of X_W beginning in $G^\sigma \times \{p_0\}$ ($0 \leqslant \sigma \leqslant \bar{\sigma}$, $p_0 \in B^{\bar{\rho}}$) is given by $(g_t, p_t) = (g_0 \exp(t\nabla w(p_0)), p_0)$. Suppose that there exists a $\nu > 0$ such that $\nabla w(p) \in \nu I_T$. In that case $t\nabla w(p_0)$ lies in $\mathfrak{t}$ (i.e., is *real*) for any $t \in \mathbb{R}$ so that $\exp(t\nabla w(p_0)) \in G$. Now $|g_0 - g_0^{\mathbb{R}}|_S \leqslant \sigma$ for some $g_0^{\mathbb{R}} \in G$ (as $g_0 \in G^\sigma$). By 6.3, we have

$$|g_0 \exp(t\nabla w(p_0)) - g_0^{\mathbb{R}} \exp(t\nabla w(p_0))|_S = |g_0 - g_0^{\mathbb{R}}|_S \leqslant \sigma \ .$$

Since $g_0^{\mathbb{R}} \exp(t\nabla w(p_0)) \in G$, this shows that $g_0 \exp(t\nabla w(p_0)) \in G^\sigma$ for all $t \in \mathbb{R}$, i.e., $(g_t, p_t) \in G^\sigma \times \{p_0\}$ for all $t \in \mathbb{R}$. Furthermore, as $\nabla w(p_0) \in \nu I_T$, this solution curve has $1/\nu$ as a period. We therefore satisfy Assumption A of Chap. 5 if we take $\mathfrak{t}^{\mathbb{Z}}$ to be any k-dimensional sublattice of I_T.

While the 'optimal' choice for $\mathfrak{t}^{\mathbb{Z}}$ is I_T itself, future computations are simplified by choosing $\mathfrak{t}^{\mathbb{Z}}$ to the sublattice generated by the inverse roots:

$$\mathfrak{t}^{\mathbb{Z}} \equiv \operatorname{span}_{\mathbb{Z}}\{\alpha_1^*, \dots, \alpha_k^*\} \ .$$

Note that $\mathfrak{t}^{\mathbb{Z}} = I_T$ if G is simply connected.

Estimating c_1

A basis $\{\beta_1, \dots, \beta_k\}$ for $\mathfrak{t}^{\mathbb{Z}}$ is given by $\beta_j \equiv \alpha_j^*$. We will estimate the constant c_1 of 5.6 in terms the Cartan integers n_{ij} of the root system. These are defined by

$$n_{ij} \equiv \frac{2B(\alpha_i, \alpha_j)}{B(\alpha_i, \alpha_i)} \in \mathbb{Z} \qquad (1 \leqslant i, j \leqslant \bar{k}) \ .$$

It is a fact that the restriction to $\mathfrak{t}$ of -1 times the Killing form can be expressed in terms of the positive real roots according to

$$a \cdot b = 8\pi^2 \sum_{m=1}^{\bar{k}} \langle \alpha_m, a \rangle \langle \alpha_m, b \rangle \qquad (a, b \in \mathfrak{t}) \ .$$

It follows from 6.11 that

6.12
$$\alpha_i^* \cdot \alpha_j^* = 8\pi^2 \sum_{m=1}^{\bar{k}} n_{im} n_{jm} \equiv m_{ij} \qquad (1 \leqslant i, j \leqslant k) \ .$$

Let $\lambda_1, \dots, \lambda_k > 0$ denote the eigenvalues of the (symmetric and positive definite) matrix $(m_{ij})_{1 \leqslant i,j \leqslant k}$. Then for some orthogonal transformation $A : \mathbb{R}^k \to \mathbb{R}^k$ and

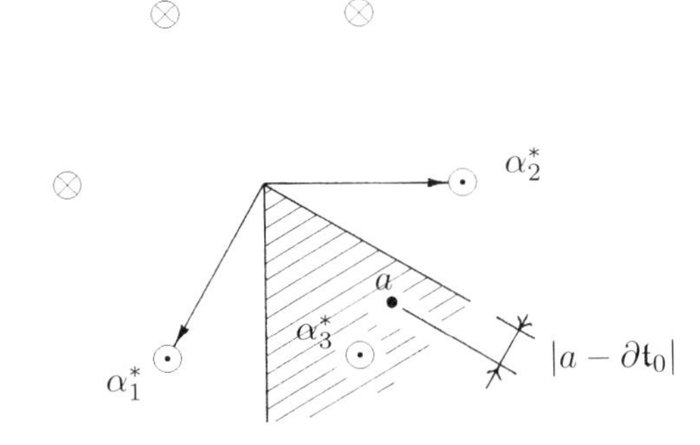

FIGURE 1. The inverse root system for $G = \mathrm{SU}(3)$ ($k = 2, \bar{k} = 3$). Here $\odot$ (resp. $\otimes$) indicates a positive (resp. negative) root. Arrows indicate basis roots. The chamber $\mathfrak{t}_0$ is shaded.

any $\tau \equiv (\tau_1, \dots, \tau_k) \in \mathbb{R}^k$,

$$\begin{aligned}|\tau_1\beta_1 + \dots + \tau_k\beta_k|^2 &= \sum_{i,j=1}^{k} \tau_i m_{ij} \tau_j = \sum_{j=1}^{k} (A\tau)_j \lambda_j (A\tau)_j \\ &\leqslant (\sup_j \lambda_j) \sum_{j=1}^{k} (A\tau)_j^2 = (\sup_j \lambda_j) \sum_{j=1}^{k} |\tau_j|^2 \\ &\leqslant (\sup_j \lambda_j) k (\sup_j |\tau_j|)^2 \ .\end{aligned}$$

We therefore satisfy 5.6 if we choose c_1 to be the square root of the maximum eigenvalue of the matrix (m_{ij}), which is defined in terms of the Cartan integers by 6.12.

6.13 Example. $G = \mathrm{SU}(3)$ (see Fig. 1). In this case $\mathfrak{t}^{\mathbb{Z}} = I_T$, $k = 2$ and $\bar{k} = 3$. The matrix of Cartan integers is

$$(n_{ij})_{1 \leqslant i,j \leqslant \bar{k}} = \begin{bmatrix} 2 & -1 & 1 \\ -1 & 2 & 1 \\ 1 & 1 & 2 \end{bmatrix} ,$$

so that

$$(m_{ij})_{1 \leqslant i,j \leqslant k} = \left(8\pi^2 \sum_{m=1}^{\bar{k}} n_{im} n_{jm} \right)_{1 \leqslant i,j \leqslant k} = 24\pi^2 \begin{bmatrix} 2 & -1 \\ -1 & 2 \end{bmatrix} .$$

The largest eigenvalue of the right-hand side is $72\pi^2$, so that for $G = \mathrm{SU}(3)$ we may take $c_1 = 6\pi\sqrt{2}$.

Bounds on λ_p

Recall from Chap. 2 that λ_p ($p \in \mathfrak{t}_0^{\mathbb{C}}$) is defined as the inverse of $\mathrm{ad}_p : \mathfrak{t}^{\perp\mathbb{C}} \to \mathfrak{t}^{\perp\mathbb{C}}$. Since λ_p appears in the equations of motion 2.10 and in the formula for the Poisson bracket 2.14, we will need to estimate the operator norm of λ_p before proceeding to verify Assumptions B and C.

6.14 Remark. Since the operator $\mathrm{ad}_p : \mathfrak{t}^{\perp\mathbb{C}} \to \mathfrak{t}^{\perp\mathbb{C}}$ becomes singular as p approaches the walls $\partial\mathfrak{t}_0$ of $\mathfrak{t}_0$, the norm of λ_p is unbounded as $p \to \partial\mathfrak{t}_0$. We will see later that this results in a deterioration in the estimates of Assumptions B and C as $\bar{p} \to \partial\mathfrak{t}_0$. This phenomenon is particular to the non-Abelian case.

Fix $p \in \mathfrak{t}_0^{\mathbb{C}}$. Note that $\dim \mathfrak{t}^{\perp\mathbb{C}} = 2\bar{k}$. For $1 \leqslant j \leqslant \bar{k}$ let ξ_j (resp. ξ_{-j}) denote an element of $\mathfrak{t}^{\perp\mathbb{C}}$ spanning the weight space corresponding to the root α_j (resp. $-\alpha_j$). Then $\{\xi_{-\bar{k}}, \dots, \xi_{-1}, \xi_1, \dots, \xi_{\bar{k}}\}$ is a basis for $\mathfrak{t}^{\perp\mathbb{C}}$ (over $\mathbb{C}$), and

$$\mathrm{ad}_p\, \xi_{\pm j} = \pm 2\pi i \langle \alpha_j, p\rangle \xi_{\pm j} \ .$$

(Here the roots α_j have been extended to $\mathbb{C}$-linear functionals on $\mathfrak{t}^{\mathbb{C}}$.) So with respect to the above basis, ad_p diagonalizes with diagonal

$$(-2\pi i\langle \alpha_{\bar{k}}, p\rangle, \dots, -2\pi i\langle \alpha_1, p\rangle, 2\pi i\langle \alpha_1, p\rangle, \dots 2\pi i\langle \alpha_{\bar{k}}, p\rangle) \ .$$

Consequently $\lambda_p = \mathrm{ad}_p^{-1}$ diagonalizes with diagonal

$$\left(\frac{-1}{2\pi i\langle \alpha_{\bar{k}}, p\rangle}, \dots, \frac{-1}{2\pi i\langle \alpha_1, p\rangle}, \frac{1}{2\pi i\langle \alpha_1, p\rangle}, \dots \frac{1}{2\pi i\langle \alpha_{\bar{k}}, p\rangle}\right) \ .$$

It follows that for any $\xi \in \mathfrak{t}^{\perp\mathbb{C}}$,

$$|\lambda_p \xi| \leqslant \frac{1}{2\pi}\left(\sup_{1\leqslant j\leqslant \bar{k}} \frac{1}{|\langle \alpha_j, p\rangle|}\right)|\xi| \ .$$

Since $\langle \alpha_j, a\rangle \in \mathbb{R}$ for all $a \in \mathfrak{t}$, $|\langle \alpha_j, p\rangle| \geqslant |\langle \alpha_j, \mathrm{Re}\, p\rangle|$. Also, every positive root α_j $(1 \leqslant j \leqslant \bar{k})$ can be written as a linear combination of the basis roots $\alpha_1, \dots, \alpha_k$ with *positive* coefficients. The above estimate can therefore be written

$$|\lambda_p \xi| \leqslant \frac{1}{2\pi}\left(\sup_{1\leqslant j\leqslant k} \frac{1}{|\langle \alpha_j, \mathrm{Re}\, p\rangle|}\right)|\xi| \qquad (\xi \in \mathfrak{t}^{\perp\mathbb{C}}) \ . \tag{6.15}$$

Manipulating 6.11, one shows that

$$\alpha_j = \frac{2\kappa(\alpha_j^*)}{|\alpha_j^*|^2} \ . \tag{6.16}$$

Recall that $p \in \mathfrak{t}_0^{\mathbb{C}}$. The diagonal matrix representation of ad_p above, Equation 2.6, and the definition of $\mathfrak{t}_0^{\mathbb{C}}$ show that

$$\mathrm{Re}\, p \in \mathfrak{t}_0 \ .$$

If $a \in \mathfrak{t}_0$, then an elementary geometric property of root systems is (see Fig. 1) that

$$|a - \partial\mathfrak{t}_0| = \inf_{1\leqslant j\leqslant k} \frac{\alpha_j^*}{|\alpha_j^*|} \cdot a \ .$$

So for any j with $1 \leqslant j \leqslant k$ one has

$$|a - \partial\mathfrak{t}_0| \leqslant \frac{\alpha_j^*}{|\alpha_j^*|} \cdot a = \frac{\langle \kappa^{-1}(\alpha_j^*), a\rangle}{|\alpha_j^*|} = \frac{1}{2}|\alpha_j^*|\langle \alpha_j, a\rangle \ ,$$

using 6.16. This shows that

$$|\langle \alpha_j, a\rangle| \geqslant \frac{2|a - \partial\mathfrak{t}_0|}{|\alpha_j^*|} \qquad (a \in \mathfrak{t}_0, 1 \leqslant j \leqslant k) \ .$$

In particular we can apply this to 6.15, with $a \equiv \operatorname{Re} p$, to conclude that

$$|\lambda_p \xi| \leqslant \frac{\sup_{1\leqslant j\leqslant k} |\alpha_j^*|}{4\pi |\operatorname{Re} p - \partial \mathfrak{t}_0|} |\xi| \ ,$$

i.e.,

$$|\lambda_p \xi| \leqslant \frac{k_3 |\xi|}{|\operatorname{Re} p - \partial \mathfrak{t}_0|} \qquad (\xi \in \mathfrak{t}^{\perp \mathbb{C}}) \ , \tag{6.17}$$

where

$$k_3 \equiv \frac{1}{4\pi} \sup_{1\leqslant j\leqslant k} |\alpha_j^*| \ . \tag{6.18}$$

Note that k_3 can be expressed purely in terms of Cartan integers: k_3^2 is the maximum diagonal entry in the matrix $(m_{ij})_{1\leqslant i,j\leqslant k}$ defined by 6.12. For example if $G = \mathrm{SU}(3)$, then by 6.13, $k_3 = 4\pi\sqrt{3}$.

Bounds on directional derivatives

In the remainder of this section the reader may find it helpful to refer frequently to the figure on p. 29. In the notation of Chap. 5, let $p_* \in \operatorname{int} B$ ($B \equiv B_{\bar\rho}(\bar p)$) and $r > 0$ be parameters subject to the constraints 5.7.

6.19 Lemma. *Let $u \in \mathcal{A}^{\mathbb{C}}(D_\gamma(p_*, r))$, $\xi \in \mathfrak{g}^{\mathbb{C}}$ and $\tau \in \mathfrak{t}^{\mathbb{C}}$ be given. If $0 < \delta \leqslant \gamma$, then for all $(g,p) \in D_{\gamma-\delta}(p_*, r)$,*

1. $$\left| \xi \cdot \frac{\partial u}{\partial g}(g,p) \right| \leqslant 2k_2(e-1) \frac{|\xi| \|u\|_\gamma^{p_*,r}}{\delta \bar\sigma} \ ,$$
2. $$\left| \tau \cdot \frac{\partial u}{\partial p}(g,p) \right| \leqslant \frac{|\tau| \|u\|_\gamma^{p_*,r}}{r \delta \bar\rho} \ .$$

PROOF. Fix $(g,p) \in D_{\gamma-\delta}(p_*, r)$. By definition

$$\xi \cdot \frac{\partial u}{\partial g}(g,p) = f'(0) \ ,$$

where $f(t) \equiv u(g \exp(t\xi), p)$. We seek $\nu > 0$ sufficiently small that $|t| \leqslant \nu$ ensures that

$$(g \exp(t\xi), p) \in D_\gamma(p_*, r) \ . \tag{6.20}$$

Then $|t| \leqslant \nu$ ensures that $f(t) \in \mathbb{C}$ is well-defined, and by Cauchy's inequality for a holomorphic function of a complex variable,

$$\left| \xi \cdot \frac{\partial u}{\partial g}(g,p) \right| \leqslant \frac{1}{\nu} \sup_{|t|\leqslant\nu} |f(t)| \leqslant \frac{\|u\|_\gamma^{p_*,r}}{\nu} \qquad (t \in \mathbb{C}) \ . \tag{6.21}$$

Since $(g,p) \in D_{\gamma-\delta}(p_*, r)$ (and because G is compact) there exists $g^{\mathbb{R}} \in G$ such that

$$|g - g^{\mathbb{R}}|_S \leqslant (\gamma - \delta)\bar\sigma \ . \tag{6.22}$$

We have

$$\begin{aligned}
|g\exp(t\xi) - G|_S &\leqslant |g\exp(t\xi) - g^{\mathbb{R}}|_S \\
&\leqslant |g\exp(t\xi) - g|_S + |g - g^{\mathbb{R}}|_S \\
&\leqslant \sqrt{n_G}|g|_S\,|\exp(t\xi) - \mathrm{id}|_S + |g - g^{\mathbb{R}}|_S \qquad \text{by 6.4} \\
&\leqslant \sqrt{n_G}\big(\,|g^{\mathbb{R}}|_S + |g - g^{\mathbb{R}}|_S\,\big)|\exp(t\xi) - \mathrm{id}|_S + |g - g^{\mathbb{R}}|_S \\
&\leqslant \sqrt{n_G}\big(1 + (\gamma - \delta)\bar{\sigma}\,\big)|\exp(t\xi) - \mathrm{id}|_S + (\gamma - \delta)\bar{\sigma} \;\; \text{by 6.2 and 6.22} \\
\text{6.23} \qquad &\leqslant 2\sqrt{n_G}|\exp(t\xi) - \mathrm{id}|_S + (\gamma - \delta)\bar{\sigma} \qquad \text{since } \bar{\sigma} \leqslant 1 \;.
\end{aligned}$$

Combining the Taylor series expansion of $\exp(t\xi) \in \mathbb{C}^{n_G \times n_G}$ with 6.4, 6.8 and 6.10:

$$\begin{aligned}
|\exp(t\xi) - \mathrm{id}|_S &\leqslant \frac{1}{\sqrt{n_G}}\left((k_2|t||\xi|) + \frac{1}{2!}(k_2|t||\xi|)^2 + \frac{1}{3!}(k_2|t||\xi|)^3 + \cdots\right) \\
&= \frac{1}{\sqrt{n_G}}(e^{k_2|t||\xi|} - 1) \\
&\leqslant \frac{k_2}{\sqrt{n_G}}(e - 1)|t||\xi| \;,
\end{aligned}$$

assuming that

$$k_2|t||\xi| \leqslant 1 \;. \tag{6.24}$$

So (assuming 6.24) 6.23 implies

$$|g\exp(t\xi) - G|_S \leqslant 2(e-1)k_2|t||\xi| + (\gamma - \delta)\bar{\sigma} \;.$$

Therefore $|g\exp(t\xi) - G|_S \leqslant \gamma\bar{\sigma}$ (so that 6.20 holds) provided

$$k_2|t||\xi| \leqslant \frac{\delta\bar{\sigma}}{2(e-1)} \;. \tag{6.25}$$

In summary, 6.20 holds if 6.24 and 6.25 do. Since 6.25 is stronger than 6.24 ($\delta, \bar{\sigma} \leqslant 1$), we may take

$$\nu \equiv \frac{\delta\bar{\sigma}}{2k_2(e-1)|\xi|} \;.$$

With this choice, 6.21 becomes 6.19.1, the first estimate of the lemma.

The proof of 6.19.2 is simpler and left to the reader. □

We write

$$\left\|\frac{\partial u}{\partial g}\right\|_{\gamma}^{p_*,r} \equiv \sup_{(g,p)\in D_\gamma(p_*,r)} \left|\frac{\partial u}{\partial g}(g,p)\right| \;.$$

One has

$$\left|\frac{\partial u}{\partial g}(g,p)\right| = \sup_{|\xi|=1}\left|\xi \cdot \frac{\partial u}{\partial g}(g,p)\right| \qquad (\xi \in \mathfrak{g}^{\mathbb{C}}) \;.$$

An analogous statement holds for $|(\partial u/\partial p)(g,p)|$. The following corollary of 6.19 is therefore clear.

6.26 Corollary. *Let $u \in \mathcal{A}^{\mathbb{C}}(D_\gamma(p_*, r))$ be given and suppose $0 < \delta \leqslant \gamma$. Then*

1. $$\left\| \frac{\partial u}{\partial g} \right\|_{\gamma-\delta}^{p_*,r} \leqslant 2k_2(e-1)\frac{\|u\|_\gamma^{p_*,r}}{\delta\bar{\sigma}} ,$$

2. $$\left\| \frac{\partial u}{\partial p} \right\|_{\gamma-\delta}^{p_*,r} \leqslant \frac{\|u\|_\gamma^{p_*,r}}{r\delta\bar{\rho}} .$$

Times of validity of Hamiltonian flows (Assumption B)

6.27 Lemma (Bounds on the components of a Hamiltonian vector field). *Let $u \in \mathcal{A}^{\mathbb{C}}(D_\gamma(p_*, r))$ be given and suppose $0 < \delta \leqslant \gamma$. Then*

1. $$\|\xi_u\|_{\gamma-\delta}^{p_*,r} \leqslant \left(1 + \frac{2k_2k_3(e-1)\bar{\rho}}{|B_{2\bar{\rho}}(\bar{p}) - \partial\mathfrak{t}_0|\bar{\sigma}}\right) \frac{\|u\|_\gamma^{p_*,r}}{r\delta\bar{\rho}} ,$$

2. $$\|\tau_u\|_{\gamma-\delta}^{p_*,r} \leqslant 2k_2(e-1)\frac{\|u\|_\gamma^{p_*,r}}{\delta\bar{\sigma}} .$$

Recall that ξ_u and τ_u are defined by 2.8 and 2.9. Note also that as $B_{2\bar{\rho}}(\bar{p})$ is closed, 6.6 ensures $|B_{2\bar{\rho}}(\bar{p}) - \partial\mathfrak{t}_0| > 0$.

PROOF. The definition 2.8 of ξ_u gives, with the help of 6.17,

$$\|\xi_u\|_{\gamma-\delta}^{p_*,r} \leqslant \left\| \frac{\partial u}{\partial p} \right\|_{\gamma-\delta}^{p_*,r} + k_3 \left(\inf_{p \in B_{r\bar{\rho}}^{(\gamma-\delta)r\bar{\rho}}(p_*)} |\operatorname{Re} p - \partial\mathfrak{t}_0| \right)^{-1} \left\| \frac{\partial u}{\partial g} \right\|_{\gamma-\delta}^{p_*,r} .$$

Recalling that 5.7 ensures that $B_{r\bar{\rho}}^{(\gamma-\delta)r\bar{\rho}} \subset B^{\bar{\rho}}$ $(0 \leqslant \gamma \leqslant 1,\ 0 < \delta \leqslant \gamma)$, we have

$$\left(\inf_{p \in B_{r\bar{\rho}}^{(\gamma-\delta)r\bar{\rho}}(p_*)} |\operatorname{Re} p - \partial\mathfrak{t}_0| \right)^{-1} \leqslant |B_{2\bar{\rho}}(\bar{p}) - \partial\mathfrak{t}_0|^{-1} .$$

Here we have used the fact that $\operatorname{Re} B^{\bar{\rho}} \subset B_{2\bar{\rho}}$. Applying 6.26, and exploiting the fact that $r \leqslant 1$, one easily deduces 6.27.1.

The inequality 6.27.2 follows similarly from 2.9 and 6.26.1. □

We are now ready to turn to the verification of Assumption B.

Let $u \in \mathcal{A}^{\mathbb{C}}(D_\gamma(p_*, r))$ be given and suppose that $0 < \delta \leqslant \gamma$. Let $t \mapsto (g_t, p_t)$ be an integral curve of X_u, and suppose that the initial condition satisfies $(g_0, p_0) \in D_{\gamma-\delta}(p_*, r)$, i.e.,

$$g_0 \in G^{(\gamma-\delta)\bar{\sigma}} \qquad p_0 \in B_{r\bar{\rho}}^{(\gamma-\delta)r\bar{\rho}}(p_*) .$$

According to the equations of motion 2.10,

$$\dot{g}_t = g_t\xi_u(g_t, p_t) , \tag{6.28}$$

$$\dot{p}_t = \tau_u(g_t, p_t) . \tag{6.29}$$

Because $D_{\gamma-\delta/2}(p_*, r)$ is compact, the solution $(g_t, p_t) \in D_{\gamma-\delta/2}(p_*, r)$ is well-defined until it reaches the boundary of $D_{\gamma-\delta/2}(p_*, r)$. That is, there exists a time st_0 $(s = \pm 1,\ 0 < t_0 \leqslant \infty)$ such that

$$|t| \leqslant t_0 \qquad \Rightarrow \qquad (g_t, p_t) \in D_{\gamma-\delta/2}(p_*, r) \tag{6.30}$$

and such that at least one of the following holds:

6.31 $$|g_{st_0} - G|_S = (\gamma - \delta/2)\bar{\sigma} \ ,$$

6.32 $$|p_{st_0} - B_{r\bar{\rho}}(p_*)| = (\gamma - \delta/2)r\bar{\rho} \ .$$

According to 6.28,

$$g_{st_0} - g_0 = \int_0^{st_0} g_t \xi_u(g_t, p_t)dt \ .$$

From 6.30 and 6.9 we deduce

$$\begin{aligned} |g_{st_0} - g_0|_S &\leqslant k_2 \left(\sup_{g \in G^{(\gamma-\delta/2)\bar{\sigma}}} |g|_S \right) \|\xi_u\|^{p_*,r}_{\gamma-\delta/2} t_0 \\ &\leqslant k_2(1 + (\gamma - \delta/2)\bar{\sigma}) \|\xi_u\|^{p_*,r}_{\gamma-\delta/2} t_0 \qquad \text{by 6.2} \\ &\leqslant 2k_2 \|\xi_u\|^{p_*,r}_{\gamma-\delta/2} t_0 \qquad (\bar{\sigma} \leqslant 1) \ . \end{aligned}$$

Applying 6.27.1:

$$|g_{st_0} - g_0|_S \leqslant 4k_2\zeta \frac{\|u\|^{p_*,r}_\gamma}{r\delta\bar{\rho}} t_0 \ ,$$

where

6.33 $$\zeta \equiv 1 + \frac{2k_2k_3(e-1)\bar{\rho}}{|B_{2\bar{\rho}}(\bar{p}) - \partial \mathfrak{t}_0|_{\bar{\sigma}}} \ .$$

6.34 Remark. Notice that $\zeta \to \infty$ as the real ball $B_{2\bar{\rho}}(\bar{p})$ approaches the walls of $\mathfrak{t}_0$.

Since $g_0 \in G^{(\gamma-\delta)\bar{\sigma}}$, there exists $g_0^{\mathbb{R}} \in G$ such that $|g_0 - g_0^{\mathbb{R}}|_S \leqslant (\gamma - \delta)\bar{\sigma}$. We compute

$$\begin{aligned} |g_{st_0} - G|_S \leqslant |g_{st_0} - g_0^{\mathbb{R}}|_S &\leqslant |g_{st_0} - g_0|_S + |g_0 - g_0^{\mathbb{R}}|_S \\ &\leqslant 4k_2\zeta \frac{\|u\|^{p_*,r}_\gamma}{r\delta\bar{\rho}} t_0 + (\gamma - \delta)\bar{\sigma} \ . \end{aligned}$$

So, supposing 6.31 holds, we have

6.35 $$t_0 \geqslant \left(\frac{1}{8k_2\zeta} \right) \frac{\bar{\sigma}\bar{\rho}\delta^2 r}{\|u\|^{p_*,r}_\gamma} \ .$$

According to 6.29

$$p_{st_0} = p_0 + \int_0^{st_0} \tau_u(g_t, p_t)dt \ .$$

From 6.30 and 6.27.2 (with δ replaced by $\delta/2$) we deduce

$$|p_{st_0} - p_0| \leqslant 4k_2(e-1)\frac{\|u\|^{p_*,r}_\gamma}{\delta\bar{\sigma}} t_0 \ .$$

Since $p_0 \in B^{(\gamma-\delta)r\bar{\rho}}_{r\bar{\rho}}(p_*)$, there exists $p_0^{\mathbb{R}} \in B_{r\bar{\rho}}(p_*)$ such that $|p_0 - p_0^{\mathbb{R}}| \leqslant (\gamma - \delta)r\bar{\rho}$. We compute

$$\begin{aligned} |p_{st_0} - B_{r\bar{\rho}}| \leqslant |p_{st_0} - p_0^{\mathbb{R}}| &\leqslant |p_{st_0} - p_0| + |p_0 - p_0^{\mathbb{R}}| \\ &\leqslant 4k_2(e-1)\frac{\|u\|^{p_*,r}_\gamma}{\delta\bar{\sigma}} t_0 + (\gamma - \delta)r\bar{\rho} \ . \end{aligned}$$

So, supposing 6.32 holds, we have

6.36
$$t_0 \geqslant \left(\frac{1}{8k_2(e-1)}\right)\frac{\bar{\sigma}\bar{\rho}\delta^2 r}{\|u\|_\gamma^{p_*,r}} .$$

Since 6.31 and/or 6.32 holds, 6.35 and/or 6.36 holds. Therefore

$$t_0 \geqslant \frac{c_2\bar{\sigma}\bar{\rho}\delta^2 r}{\|u\|_\gamma^{p_*,r}} ,$$

if we take

6.37
$$c_2 \equiv \frac{1}{8k_2 \max\{\zeta, e-1\}} .$$

Arguments like the above also show that

$$|p_t - p_0| \leqslant \left(\frac{|t|}{t_0}\right)\left(\frac{\delta}{2}\right) r\bar{\rho} \qquad (|t| \leqslant t_0) .$$

With the above choice of c_2, Assumption B of Chap. 5 is thus established.

Inequalities for Poisson brackets (Assumption C)

Let $u, v \in \mathcal{A}^{\mathbb{C}}(D_\gamma(p_*, r))$ be given and suppose $0 < \delta \leqslant \gamma$. Fixing some $(g,p) \in D_{\gamma-\delta}(p_*, r)$, we have

$$\begin{aligned}\{u,v\}(g,p) &= \langle du, X_v(g,p)\rangle \\ &= \xi_v(g,p)\cdot\frac{\partial u}{\partial g}(g,p) + \tau_v(g,p)\cdot\frac{\partial u}{\partial p}(g,p) \qquad \text{by 2.7} \\ &= f'(0) ,\end{aligned}$$

where

$$f(t) \equiv u(g\exp(t\xi_v(g,p)), p + t\tau_v(g,p)) .$$

We seek $\nu > 0$ sufficiently small that $|t| \leqslant \nu$ guarantees

6.38
$$(g\exp(t\xi_v(g,p)), p + t\tau_v(g,p)) \in D_\gamma(p_*, r) .$$

Then $|t| \leqslant \nu$ ensures that $f(t) \in \mathbb{C}$ is well-defined and, by Cauchy's inequality,

6.39
$$|\{u,v\}(g,p)| \leqslant \frac{1}{\nu}\sup_{|t|\leqslant\nu}|f(t)| \leqslant \frac{\|u\|_\gamma^{p_*,r}}{\nu} .$$

To satisfy 6.38 we need

6.40
$$|g\exp(t\xi_v(g,p)) - G|_S \leqslant \gamma\bar{\sigma}$$

6.41
$$\text{and} \qquad |(p + t\tau_v(g,p)) - B_{r\bar{\rho}}(p_*)| \leqslant \gamma r\bar{\rho} .$$

Arguing as in the proof of 6.19.1, we satisfy 6.40 if

$$|t| \leqslant \frac{\delta\bar{\sigma}}{2k_2(e-1)\,|\xi_v(g,p)|} .$$

Since $(g,p) \in D_{\gamma-\delta}(p_*, r)$, it suffices, by 6.27.1, to ensure

6.42
$$|t| \leqslant \frac{\bar{\sigma}\bar{\rho}\delta^2 r}{2k_2(e-1)\zeta\,\|v\|_\gamma^{p_*,r}} ,$$

where ζ is the constant defined by 6.33.

Arguing similarly, one sees that 6.41 holds provided

$$|t| \leqslant \frac{r\delta\bar{\rho}}{|\tau_v(g,p)|} \ .$$

By 6.27.2, it suffices to ensure

6.43
$$|t| \leqslant \frac{\bar{\sigma}\bar{\rho}\delta^2 r}{2k_2(e-1)\,\|v\|_\gamma^{p_*,r}} \ .$$

To summarize, if 6.42 and 6.43 hold, then so do 6.40 and 6.41, and hence 6.38. It follows that we may take

$$\nu \equiv \frac{\bar{\sigma}\bar{\rho}\delta^2 r}{2k_2(e-1)\max\{1,\zeta\}\,\|v\|_\gamma^{p_*,r}} \ .$$

With this choice of ν, 6.39 becomes

$$|\{u,v\}(g,p)| \leqslant \left(\frac{2k_2(e-1)\max\{1,\zeta\}}{\bar{\sigma}\bar{\rho}}\right)\frac{\|u\|_\gamma^{p_*,r}\,\|v\|_\gamma^{p_*,r}}{\delta^2 r} \ .$$

Since $(g,p) \in D_{\gamma-\delta}(p_*,r)$ was arbitrary, the first estimate of Assumption C holds if we take $c_3 \equiv 2k_2(e-1)\max\{1,\zeta\}$.

If $W \in \mathcal{A}^{\mathbb{C}}(D_\gamma(p_*,r))$ is of the form $W(g,p) = w(p)$, then (by, e.g., 2.14),

$$\{u,W\}(g,p) = \frac{\partial u}{\partial g}(g,p)\cdot\nabla w(p) = f'(0) \qquad (\,(g,p) \in D_{\gamma-\delta}(p_*,r)\,) \ ,$$

where $f(t) \equiv u(g\exp(t\nabla k(p)),p)$. Cauchy's inequality and the by now familiar argument leads to the estimate

$$\|\{u,W\}\|_\gamma^{p_*,r} \leqslant \frac{2k_2(e-1)}{\delta\bar{\sigma}} \sup_{p\in B_{r\bar{\rho}}^{(\gamma-\delta)r\bar{\rho}}(p_*)} |\nabla w(p)| \ .$$

Since $\delta \leqslant 1$, the second estimate of Assumption C certainly holds if we take

6.44
$$c_4 \equiv 2k_2(e-1) \ .$$

This completes the proof of Proposition 6.1 in the case of a semisimple G.

CHAPTER 7

Nekhoroshev-Type Estimates for Momentum Maps

In this section we combine Theorem 3.10, Theorem 5.8 and Proposition 6.1 to obtain the Nekhoroshev type estimates on momentum maps stated in the Introduction.

Let G be a compact connected Lie group and $(P, \omega, G, \mathbf{J})$ a Hamiltonian G-space. Consider a Hamiltonian of the form

$$H = H_0 + F \ ,$$

where H_0 is G-invariant, and F is arbitrary.

Existence of action-group coordinates. Assume that the hypotheses of Theorem 3.10 apply (working 'locally' à la Remark 3.11 if necessary), so that there exists a diffeomorphism $\phi : G \times U \to P$ $(U \equiv \varphi^{-1}(\mathbf{J}(P)) \cap \mathfrak{t}_0)$ such that $(H_0 \circ \phi)(g, p) = h(p)$, for some smooth $h : U \to \mathbb{R}$. Recall that $\varphi : \mathfrak{g} \to \mathfrak{g}^*$ is the isomorphism corresponding to the fixed Ad-invariant inner product on $\mathfrak{g}$.

Complexification of the coordinates. Fix some $\bar{x} \in P$ and define $(\bar{g}, \bar{p}) \equiv \phi^{-1}(\bar{x})$. Fix a positive constant $\bar{\rho}$ small enough that the real closed ball $B_{2\bar{\rho}}(\bar{p})$ is contained in $U \subset \mathfrak{t}_0$ and the complex closed ball $B_{\bar{\rho}}^{\bar{\rho}}(\bar{p})$ (defined in Chap. 5) is contained in $\mathfrak{t}_0^{\mathbb{C}}$ (defined in Chap. 2). Fix a positive constant $\bar{\sigma} \leqslant 1$ small enough that $(G^\sigma)_{0 \leqslant \sigma \leqslant \bar{\sigma}}$, as defined by 6.5, is a local complexification of G in the sense of 5.1.

Analyticity of the Hamiltonian. Assume that $h \in \mathcal{A}^{\mathbb{R}}(B_{\bar{\rho}}(\bar{p}))$ and $F \circ \phi \in \mathcal{A}^{\mathbb{R}}(G \times B_{\bar{\rho}}(\bar{p}))$ (notation as in Chap. 5). Furthermore assume that $\bar{\sigma}$ and $\bar{\rho}$ are sufficiently small that h and $F \circ \phi$ have holomorphic extensions $h \in \mathcal{A}^{\mathbb{C}}(B_{\bar{\rho}}^{\bar{\rho}}(\bar{p}))$ and $F \circ \phi \in \mathcal{A}^{\mathbb{C}}(G^{\bar{\sigma}} \times B_{\bar{\rho}}^{\bar{\rho}}(\bar{p}))$.

Convexity of the unperturbed Hamiltonian. Assume that h is (m, M)-convex on int $B_{\bar{\rho}}^{\bar{\rho}}(\bar{p})$ in the sense of 5.3, for some $m, M > 0$.

Constants. Write $\Omega(p) = \nabla h(p) \in \mathfrak{t}$ and define the 'time scales'

$$T_m \equiv \frac{1}{\bar{\rho} m} \qquad T_M \equiv \frac{1}{\bar{\rho} M}$$

$$T_\Omega \equiv \left(\sup_{p \in B^{\bar{\rho}}} |\Omega(p)| \right)^{-1} \qquad T_{h'''} \equiv \left(\bar{\rho}^2 \sup_{p \in B} \sup_{u,v,w \in \mathfrak{t}} |D^3 h(p)(u, v, w)| \right)^{-1}$$

and the associated 'dimensionless constants'

$$c_5 \equiv \frac{T_{h'''}}{T_m} \qquad c_6 \equiv \frac{T_M}{T_\Omega} \qquad c_7 \equiv \frac{T_M}{T_m} = \frac{m}{M} \ .$$

Recall that we allow $T_{h'''} = \infty$.

The perturbation parameter. Let $\|\cdot\|$ denote the supremum norm on $\mathcal{A}^{\mathbb{C}}(G^{\bar{\sigma}} \times B^{\bar{\rho}})$ and define the 'perturbation parameter'

$$\epsilon \equiv \frac{\|F\|}{E} ,$$

where $E \equiv \bar{\sigma}\bar{\rho}/T_M$. We have the following corollary of Theorem 5.8 and Proposition 6.1. (Recall that k denotes the rank of G.)

7.1 Corollary. *There exist positive constants c_1, c_2, c_3, c_4 independent of h and the perturbation F, and positive constants $a = a(k)$, $b = b(k)$, $c = c(c_1, \dots, c_7)$, $\epsilon_0 = \epsilon_0(k, c_1, \dots, c_7)$, $t_0 = t_0(c_1, \dots, c_7)$ and $r_0 = r_0(c_1, \dots, c_7)$, such that for all $\epsilon \leqslant \epsilon_0$ and every integral curve $t \mapsto x_t \in P$ of X_H with $x_0 = \bar{x}$ one has*

$$\frac{|t|}{T_\Omega} \leqslant t_0 \exp(c\epsilon^{-a}) \ \Rightarrow\ \frac{|\mathbf{J}(x_t) - \mathcal{O}|}{\bar{\rho}} \leqslant r_0 \epsilon^b ,$$

where $\mathcal{O} \subset \mathfrak{g}^*$ is the co-adjoint orbit through the point $\mathbf{J}(x_0)$.

Estimates for the constants $a, b, c, \epsilon_0, t_0, r_0$ appear in Addendum 5.9. If G is semisimple (and one takes -1 times the Killing form as the fixed inner product on $\mathfrak{g}$), then c_1 depends only on the Cartan integers of G (according to the explicit computation presented on p. 33) and one may take

$$\begin{aligned} c_2 &\equiv \frac{1}{8k_2 \max\{\zeta, e-1\}} \\ c_3 &\equiv 2k_2(e-1)\max\{1, \zeta\} \\ c_4 &\equiv 2k_2(e-1) \\ \zeta &\equiv 1 + \frac{2k_2k_3(e-1)\bar{\rho}}{|B_{2\bar{\rho}}(\bar{p}) - \partial\mathfrak{t}_0|\bar{\sigma}} . \end{aligned}$$

The constant k_3 (see 6.18) depends only on the Cartan integers. The constant k_2 (see 6.10) depends on the realization of G as a linear group.

The reader should keep Remark 6.34 in mind.

PROOF. That one arrives at the following estimate for the action variables is clear:

$$\frac{|t|}{T_\Omega} \leqslant \exp(c\epsilon^{-a}) \ \Rightarrow\ \frac{|p_t - p_0|}{\bar{\rho}} \leqslant r_0\epsilon^b \qquad \big((g_t, p_t) \equiv \phi^{-1}(x_t)\big) .$$

According to Theorem 3.10 and 3.3, we have $\mathbf{J}(x_t) = (\mathbf{J}^G \circ \phi^{-1})(x_t) = g_t \cdot \varphi^{-1}(p_t)$ and compute

$$\begin{aligned} |\mathbf{J}(x_t) - \mathcal{O}| &\leqslant \big|\mathbf{J}(x_t) - g_t g_0^{-1} \cdot \mathbf{J}(x_0)\big| = \big|g_t \cdot \varphi^{-1}(p_t) - g_t \cdot \varphi^{-1}(p_0)\big| \\ &= \big|\varphi^{-1}(p_t) - \varphi^{-1}(p_0)\big| = |p_t - p_0| , \end{aligned}$$

since the isomorphism $\varphi : \mathfrak{t} \to \underline{\mathfrak{t}}$ is isometric. This proves that the estimate claimed in the corollary indeed holds. □

Part 2

Geometry

CHAPTER 8

On Hamiltonian G-Spaces with Regular Momenta

In this chapter we assume some familiarity with the structure of compact Lie groups, say with what has been summarized in Chap. 1 of Part 1. Notation borrowed from Part 1 may be deciphered using the index on p. ix. We work throughout in the smooth category, where 'smooth' means C^∞ or real-analytic.

Preliminaries

Let $(P, \omega, G, \mathbf{J})$ be a Hamiltonian G-space, as defined on p. 15. Recall that the *co-adjoint orbit type* of a point $x \in P$ is the orbit type of $\mathbf{J}(x) \in \mathfrak{g}^*$. The space P has uniform co-adjoint orbit type (T) ($T \subset G$ denoting a maximal torus) if and only if $\mathbf{J}(P) \subset \mathfrak{g}^*_{\mathrm{reg}}$. In that case we will call P a *Hamiltonian G-space with regular momenta.* Our interest in such spaces is motivated by the fact that, under an appropriate integrability hypothesis, they admit action-group coordinates. A simple example is geodesic motions on S^2; see Example 3.4 and below.

For us the construction of 'coordinates' in a Hamiltonian system will mean the construction of an appropriate *equivalence* between Hamiltonian G-spaces:

8.1 Definition. Two Hamiltonian G-spaces $(P_1, \omega_1, G, \mathbf{J}^1)$ and $(P_2, \omega_2, G, \mathbf{J}^2)$ will be said to be *equivalent* if there exists a G-equivariant symplectic diffeomorphism $\phi : P_1 \to P_2$ such that $\mathbf{J}^2 \circ \phi = \mathbf{J}^1$.

> *Henceforth $T \subset G$ denotes a fixed maximal torus, $\mathfrak{t} \subset \mathfrak{g}$ its Lie algebra, and $\mathcal{W} \subset \underline{\mathfrak{t}} \equiv \mathrm{Ann}[\mathfrak{g}, \mathfrak{t}]$ a fixed (open) Weyl chamber in $\mathfrak{g}^*$; $\mathcal{O}$ will denote a fixed regular co-adjoint orbit, and μ_0 the unique point of intersection of $\mathcal{O}$ with $\mathcal{W}$.*

The natural projection $\mathfrak{g}^* \to \mathfrak{t}^*$ restricts to an isomorphism $i : \underline{\mathfrak{t}} \to \mathfrak{t}^*$, which identifies $\mathcal{W}$ with an open set $\mathfrak{t}^*_0 \equiv i(\mathcal{W})$. We refer to $\mathfrak{t}^*_0$ as a Weyl chamber also.

We now turn to a systematic study of Hamiltonian G-spaces with regular momenta. While we make no explicit assumptions on the orbit type (G_x) of points $x \in P$, the following is worth noting.

8.2 Proposition. *If P is a Hamiltonian G-space with regular momenta, then $\dim G_x \leqslant \operatorname{rank} G$ for all $x \in P$.*

This follows from the equivariance of $\mathbf{J}$ and the identity

$$\mathfrak{g}_x = \mathrm{Ann}\, \mathrm{Im}\, \mathrm{T}_x \mathbf{J} \ , \tag{8.3}$$

which follows from the definition of a momentum map.

For any Hamiltonian G-space P, the open G-invariant subset $P' \equiv \mathbf{J}^{-1}(\mathfrak{g}^*_{\mathrm{reg}})$ is a Hamiltonian G-space with regular momenta. If, however, $\dim G_x > \operatorname{rank} G$ for all $x \in P$, then Proposition 8.2 implies that P' is empty. The following example (relevant to the so-called 1:1:1 resonance) is a case in point.

8.4 Example. Equip $P \equiv \mathbb{R}^6$ with the standard symplectic structure, and let $G \equiv \mathrm{SU}(3)$ act (faithfully) on P by identifying $\mathbb{R}^6$ with $\mathbb{C}^3$. Then G acts symplectically, as is well-known, and so admits a momentum map, on account of the Poincaré lemma. This momentum map is equivariant since SU(3) is semisimple (Guillemin and Sternberg, 1984). One finds that for all $x \in P$, dim $G_x \geq 3 >$rank $G = 2$, and that accordingly $\mathbf{J}(P) \cap \mathfrak{g}^*_{\mathrm{reg}} = \emptyset$. (In the 1:1:1 resonance one studies cubic and higher order perturbations of the G-invariant Hamiltonian $H(z_1, z_2, z_3) = \frac{1}{2}|z_1|^2 + \frac{1}{2}|z_2|^2 + \frac{1}{2}|z_3|^2$.)

Henceforth $(P, \omega, G, \mathbf{J})$ denotes a Hamiltonian G-space with $\mathbf{J}(P) \subset \mathfrak{g}^*_{\mathrm{reg}}$.

On the geometry of the co-adjoint action

Our first observation is that there exists a natural G-equivariant trivialization $\mathfrak{g}^*_{\mathrm{reg}} \xrightarrow{\sim} \mathcal{O} \times \mathcal{W}$ (with G acting on $\mathcal{O} \times \mathcal{W}$ according to $g \cdot (\nu, w) \equiv (g \cdot \nu, w)$). This trivialization is given by $\mu \mapsto (\pi_{\mathcal{O}}(\mu), \pi_{\mathcal{W}}(\mu))$, where $\pi_{\mathcal{O}} : \mathfrak{g}^*_{\mathrm{reg}} \to \mathcal{O}$ and $\pi_{\mathcal{W}} : \mathfrak{g}^*_{\mathrm{reg}} \to \mathcal{W}$ are defined implicitly by

8.5
$$\begin{aligned} \pi_{\mathcal{O}}(g \cdot w) &\equiv g \cdot \mu_0 \\ \pi_{\mathcal{W}}(g \cdot w) &\equiv w \end{aligned} \qquad (g \in G, w \in \mathcal{W}) \ ,$$

every element of $\mathfrak{g}^*_{\mathrm{reg}}$ being of the form $g \cdot w$ for some $g \in G$ and $w \in \mathcal{W}$. Indeed $\mathcal{W}$ is a global slice at μ_0 for the co-adjoint action of G on $\mathfrak{g}^*_{\mathrm{reg}}$, so that there exists a natural G-equivariant isomorphism $\mathfrak{g}^*_{\mathrm{reg}} \cong G \times_T \mathcal{W}$ (see Appendix B). Since all points in $\mathcal{W}$ have isotropy group T, we have $G \times_T \mathcal{W} \cong G/T \times \mathcal{W} \cong \mathcal{O} \times \mathcal{W}$.

Note that the fibers of $\pi_{\mathcal{O}} : \mathfrak{g}^*_{\mathrm{reg}} \to \mathcal{O}$ are the Weyl chambers in $\mathfrak{g}^*$; the fibers of $\pi_{\mathcal{W}} : \mathfrak{g}^*_{\mathrm{reg}} \to \mathcal{W}$ are the regular co-adjoint orbits.

8.6 Example. Take $G = \mathrm{SO}(3)$ and let $T \subset G$ be the rotations about the e_3 axis. The Lie algebra $\mathfrak{g} = \mathfrak{so}(3)$ can be identified with $\mathbb{R}^3$ via the isomorphism $\xi \mapsto \hat{\xi} : \mathbb{R}^3 \to \mathfrak{so}(3)$ defined by $\hat{\xi} u = \xi \times u$ $(u \in \mathbb{R}^3)$. We have a corresponding identification $\mathfrak{g}^* \cong \mathbb{R}^3$. The co-adjoint action is then represented by the standard action of SO(3) on $\mathbb{R}^3$. We have $\underline{\mathfrak{t}} = \mathrm{span}\{e_3\}$. For a Weyl chamber in $\mathfrak{g}^*$ choose $\mathcal{W} \equiv \{te_3 \mid t > 0\}$. For a regular co-adjoint orbit $\mathcal{O}$ choose the unit sphere $S^2 \subset \mathbb{R}^3$. The unique intersection point of $\mathcal{W}$ and $\mathcal{O}$ is $\mu_0 \equiv e_3$. The projection $\pi_{\mathcal{O}} : \mathbb{R}^3 \backslash \{0\} \to S^2$ is given by $\pi_{\mathcal{O}}(\mu) \equiv \mu / \|\mu\|$. The projection $\pi_{\mathcal{W}} : \mathbb{R}^3 \backslash \{0\} \to \mathcal{W} \cong (0, \infty)$ is given by $\pi_{\mathcal{W}}(\mu) \equiv \|\mu\|$.

Since $\mathbf{J}(P) \subset \mathfrak{g}^*_{\mathrm{reg}}$, it is natural to pull the projections $\pi_{\mathcal{O}}$ and $\pi_{\mathcal{W}}$ back using the momentum map: Define

8.7
$$\begin{aligned} \pi &\equiv \pi_{\mathcal{O}} \circ \mathbf{J} : P \to \mathcal{O} \ , \\ \mathbf{j} &\equiv i \circ \pi_{\mathcal{W}} \circ \mathbf{J} : P \to \mathfrak{t}^*_0 \subset \mathfrak{t}^* \ . \end{aligned}$$

In the following paragraphs we shall see that each fiber of π encodes the geometric structure of the full space P. Later, we shall see that $\mathbf{j} : P \to \mathfrak{t}^*$ is the momentum map for a natural Hamiltonian action of T on P. Unlike the action of T as a subgroup of G, this new T-action leaves the fibers of π invariant. Furthermore the orbits of this action do not depend on the choice of maximal torus T.

The fibering of P by Hamiltonian T-spaces

The map π is G-equivariant. Since G acts transitively on $\mathcal{O}$, it follows that $\pi : P \to \mathcal{O}$ is a surjective submersion.

8.8 Theorem (Guillemin-Sternberg). *Each fiber $F_\nu \equiv \pi^{-1}(\nu)$ of π is a symplectic submanifold of P invariant with respect to the action of the maximal torus G_ν. Furthermore, G_ν acts on F_ν in a Hamiltonian fashion (with respect to the symplectic structure $\omega_\nu \equiv \omega|F_\nu$) with an equivariant momentum map $\mathbf{J}^\nu : F_\nu \to \mathfrak{g}_\nu^*$ given by $\mathbf{J}^\nu(x) \equiv \mathbf{J}(x)|\mathfrak{g}_\nu$ ($x \in F_\nu$). This makes F_ν a Hamiltonian G_ν-space $(F_\nu, \omega_\nu, G_\nu, \mathbf{J}^\nu)$.*

Each element $g \in G$ maps each fiber of π symplectomorphically onto another fiber, and this induced action on the fibers is transitive.

Theorem 8.8 follows, for example, from the symplectic cross-section theorem (Guillemin and Sternberg, 1984, §41). For a statement of this theorem the reader is referred to Guillemin et al. (1996), from whom we also borrow the following terminology:

8.9 Definition. Write $F \equiv F_{\mu_0}$, $\omega_F \equiv \omega_{\mu_0}$ and $\mathbf{J}^F \equiv \mathbf{J}^{\mu_0}$. The Hamiltonian T-space $(F, \omega_F, T, \mathbf{J}^F)$ will be referred to as the *symplectic cross-section* of $(P, \omega, G, \mathbf{J})$. Note that $\mathbf{J}^F = i \circ \mathbf{J}|F$.

In fact, the symplectic cross-section theorem gives more information than is contained in the statement of Theorem 8.8. It shows that $\pi : P \to \mathcal{O}$ is naturally isomorphic to the bundle $G \times_T F \to G/T$. In particular, $\pi : P \to \mathcal{O}$ is a *symplectic fibration*, meaning that it is a locally trivial fiber bundle, with fibers modeled on F, whose structure group is contained in the symplectomorphism group of F. The symplectic cross-section theorem also gives a characterization of the symplectic connection on $\pi : P \to \mathcal{O}$ (see 8.11 below for a definition). This characterization allows one to recover the symplectic structure ω on P from: (i) its restriction to F, and (ii) the natural co-adjoint orbit symplectic structure of the base $\mathcal{O}$. The interested reader is referred to op. cit. for details. For our purposes, Theorem 8.8 and the rather straightforward observation in Lemma 8.12 below will suffice.

Since all maximal tori of G are conjugate, we can replace the action of G_ν on a fiber F_ν ($\nu \in \mathcal{O}$) by a Hamiltonian action of the fixed torus $T = G_{\mu_0}$. (In fact, a slightly stronger statement is true; see Corollary 8.18.) This makes each fiber a Hamiltonian T-space. By Theorem 8.8 there exists for any fibers F_{ν_1} and F_{ν_2} an element of G mapping F_{ν_1} symplectomorphically onto F_{ν_2}. This map is in fact an *equivalence* of Hamiltonian T-spaces, in the sense of 8.1. In particular, each fiber of π is equivalent as a Hamiltonian T-space to the symplectic cross-section.

8.10 Remark (On G-invariant dynamics). If $H : P \to \mathbb{R}$ is a G-invariant Hamiltonian, then the fibers of $\pi \equiv \pi_\mathcal{O} \circ \mathbf{J}$ will be X_H-invariant, by Noether's theorem (which states that the preimage of *any* set under the momentum map is X_H-invariant). Furthermore, the dynamics in any two fibers will be conjugate because any fiber can be mapped symplectomorphically onto any other by a group element. So any feature of the dynamics in a given fiber is repeated in *all* the fibers. For example, the existence of an equilibrium point for X_H in fact implies the existence of an entire manifold of equilibria diffeomorphic to $\mathcal{O}$. Borrowing language from perturbation theory (where one is interested in non-invariant perturbations to H), we say that such a Hamiltonian is more 'resonant.'[1] This built-in resonance is particular to the non-Abelian case, for otherwise $\mathcal{O}$ is just a point and π has only one fiber, namely P.

The symplectic connection

8.11 Definition. The *symplectic connection* on $\pi : P \to \mathcal{O}$ is the assignment to each $x \in P$ of a space $\mathrm{hor}_x \equiv (\ker \mathrm{T}_x\pi)^\omega$ called the *horizontal space* at x which, on account of $F_{\pi(x)} \subset P$ being a symplectic submanifold, is complementary to $\ker \mathrm{T}_x\pi$ in T_xP.

The restriction of ω to the horizontal spaces is easily characterized:

8.12 Lemma. 1. $\mathrm{hor}_x \subset \mathrm{T}_x(G \cdot x) \qquad (x \in p)$.
2. $\omega(\xi_P(x), \eta_P(x)) = \langle \mathbf{J}(x), [\xi, \eta] \rangle \qquad (x \in P;\ \xi, \eta \in \mathfrak{g})$.

PROOF. Since $\pi = \pi_\mathcal{O} \circ \mathbf{J}$, we have $\ker \mathrm{T}_x\pi \supset \ker \mathrm{T}_x\mathbf{J}$. Taking ω-orthogonal complements, we obtain $\mathrm{hor}_x \subset (\ker \mathrm{T}_x\mathbf{J})^\omega$. By the definition of a momentum map, $\mathfrak{g}^* \xleftarrow{\mathbf{J}} P \to P/G$ has the 'dual pair' property, i.e. $(\ker \mathrm{T}_x\mathbf{J})^\omega = \mathrm{T}_x(G \cdot x)$. So 1 holds.

To prove 2, we observe that as $\mathbf{J}$ is equivariant, it is also infinitesimally equivariant, from which it follows that $\{J_\xi, J_\eta\} = J_{[\xi,\eta]}$, where $\{f, h\} \equiv X_h \lrcorner X_f \lrcorner \omega$ (see, e.g., Abraham and Marsden (1978, Theorem 4.2.8) for a proof). We can now compute

$$\begin{aligned} \omega(\xi_P(x), \eta_P(x)) &= \omega(X_{J_\xi}(x), X_{J_\eta}(x)) = \{J_\xi, J_\eta\}(x) \\ &= J_{[\xi,\eta]}(x) = \langle \mathbf{J}(x), [\xi, \eta] \rangle \ . \end{aligned}$$

□

A simple example: Geodesic motions on the sphere

We now describe a simple example for which one can give an explicit realization of the symplectic cross-section.

Recall (see 3.4, p. 17) that the phase space for geodesic motions on S^2 can be identified with a Hamiltonian G-space $(\mathrm{T}S^2, \omega, \mathrm{SO}(3), \mathbf{J})$ where $\mathbf{J} : \mathrm{T}S^2 \to \mathfrak{so}(3)^* \cong \mathbb{R}^3$ is given by $\mathbf{J}(q, v) \equiv q \times v$. (Here we are viewing $\mathrm{T}S^2$ as $\{(q, v) \in S^2 \times \mathbb{R}^3 \mid q \cdot v = 0\}$.) If we restrict attention to the open dense invariant subset $P \subset \mathrm{T}S^2$ obtained by removing the zero section, we obtain a Hamiltonian G-space $(P, \omega, \mathrm{SO}(3), \mathbf{J})$ with regular momenta.

Choose T, $\mathcal{W}$ and $\mathcal{O}$ as in Example 8.6. Then the symplectic fibration $\pi : P \to S^2$ is given by $\pi(q, v) \equiv \pi_\mathcal{O}(\mathbf{J}(q, v)) = v / \|v\|$.

The symplectic cross-section $F = \pi^{-1}(e_3)$ is symplectomorphic to the cylinder $S^1 \times (0, \infty)$, equipped with the standard symplectic structure $d\theta \wedge dI$. Indeed a diffeomorphism $\varphi : S^1 \times (0, \infty) \to F$ is given by

$$\varphi(\theta, I) \equiv (R^3_\theta e_1, I R^3_\theta e_2) \ ,$$

where $R^3_\theta \equiv \exp(\theta \hat{e}_3)$. To show that φ is symplectic, recall that $\omega = -d\Theta$ where

8.13 $$\langle \Theta, \frac{d}{dt}(q_t, v_t)\big|_{t=0} \rangle \equiv v_0 \cdot \dot{q}_0 \ .$$

---

[1]A perturbation theorist might alternatively use the term 'degenerate.' From the symmetry point of view, this terminology is misleading, as the phenomenon we have just described applies to *any* G-invariant Hamiltonian.

From this one computes $\frac{\partial}{\partial\theta} \lrcorner \varphi^*\Theta = I$ and $\frac{\partial}{\partial I} \lrcorner \varphi^*\Theta = 0$, so that $\varphi^*\Theta = I d\theta$, giving $\varphi^*\omega = d\theta \wedge dI$ as required.

The action of $T \cong S^1$ on $S^1 \times (0,\infty)$ that makes the symplectomorphism φ a T-equivariant map is given by $\theta' \cdot (\theta, I) \equiv (\theta + \theta', I)$. Identify the Lie algebra of S^1 (and its dual) with $\mathbb{R}$ using the generator $\frac{\partial}{\partial\theta}(0)$. Then the momentum map $\mathbf{J}^F \circ \varphi$ of this action is given by the coordinate function $I : S^1 \times (0,\infty) \to \mathbb{R}$.

To summarize, the symplectic cross-section of the Hamiltonian G-space applying to geodesic motions on S^2 is just the cylinder $S^1 \times (0,\infty)$ equipped with the standard S^1 Hamiltonian action.

The Hamiltonian for geodesic motions on S^2 is (up to a rescaling) $H(q,v) = \frac{1}{2}\|v\|^2$, which is SO(3)-invariant. This Hamiltonian restricts to a function on F that is pulled back by $\varphi : S^1 \times (0,\infty) \xrightarrow{\sim} F$ to $\frac{1}{2}I^2$. This is the Hamiltonian for geodesic motions on the *circle*, and is S^1-invariant. Thus geodesic motions on the sphere can be imagined as a family of subsystems F_ν $(\nu \in \mathcal{O})$, each identifiable with geodesic motions on the circle. This corresponds precisely to the following familiar fact: A point mass constrained to move on the surface of a smooth sphere moves as if it were constrained to move on a great circle. This circle is given by the intersection of the sphere with the plane whose normal is aligned with the initial (and subsequently conserved) angular momentum. These normals live in the space $S^2 = \mathcal{O}$, the 'unit momentum sphere.'

Reduction to the Abelian case

Here is the central result of this chapter.

8.14 Theorem.
Two Hamiltonian G-spaces $(P_1, \omega_1, G, \mathbf{J}^1)$ and $(P_2, \omega_2, G, \mathbf{J}^2)$ with regular momenta are equivalent (in the sense of 8.1) if and only if their symplectic cross-sections are equivalent. Explicitly, if $\varphi : F_1 \to F_2$ is an equivalence between the symplectic cross-sections, then a well-defined equivalence $\phi : P_1 \to P_2$ is given by

$$\phi(g \cdot y) \equiv g \cdot \varphi(y) \qquad (g \in G, y \in F_1) \ .$$

PROOF. Recall that as G acts transitively on the fibers of $\pi_1 : P_1 \to \mathcal{O}$ (the symplectic fibration associated with $(P_1, \omega_1, G, \mathbf{J}^1)$), every element of P_1 is of the form $g \cdot y$ for some $g \in G$ and $y \in F_1 = \pi_1^{-1}(\mu_0)$. Furthermore, $g \cdot y = g' \cdot y'$ $(g' \in G,\ y' \in F_1)$ if and only if $g' = gq$ and $y' = q^{-1} \cdot y$, for some $q \in T$ (by the G-equivariance of π_1). In that case $g' \cdot \varphi(y') = gq \cdot \varphi(q^{-1} \cdot y) = g \cdot \varphi(y)$, by the T-equivariance of φ. This shows that ϕ is well-defined.

That ϕ is smooth is clear. It is a diffeomorphism since it has a well-defined and smooth inverse $\phi^{-1} : P_2 \to P_1$ given by

$$\phi^{-1}(g \cdot y) \equiv g \cdot \varphi^{-1}(y) \qquad (g \in G, y \in F_2) \ .$$

By construction ϕ is G-equivariant.

We show next that $\mathbf{J}^2 \circ \phi = \mathbf{J}^1$. Let $g \cdot y \in P_1$ be arbitrary $(g \in G,\ y \in F_1)$. Then $\mathbf{J}^1(y) \in \mathcal{W} \subset \underline{\mathfrak{t}}$, so that $\mathbf{J}^1(y) = i^{-1}(\mathbf{J}^1(y)|\mathfrak{t}) = i^{-1}(\mathbf{J}^{F_1}(y))$. Similarly, as $\varphi(y) \in F_2$, we have $\mathbf{J}^2(\varphi(y)) = i^{-1}(\mathbf{J}^{F_2}(\varphi(y)))$. But $\mathbf{J}^{F_2} \circ \varphi = \mathbf{J}^{F_1}$, since $\varphi : F_1 \to F_2$ is an equivalence, from which we deduce

$$\mathbf{J}^2(\varphi(y)) = \mathbf{J}^1(y) \ . \tag{8.15}$$

Using this fact we now compute

$$(\mathbf{J}^2 \circ \phi)(g \cdot y) = \mathbf{J}^2(g \cdot \varphi(y)) = g \cdot \mathbf{J}^2(\varphi(y)) = g \cdot \mathbf{J}^1(y) = \mathbf{J}^1(g \cdot y) \ .$$

Since $g \cdot y \in F_1$ was arbitrary, this shows that $\mathbf{J}^2 \circ \phi = \mathbf{J}^1$.

To prove that ϕ is an equivalence it remains to show that ϕ is symplectic. Since $\varphi : F_1 \longrightarrow F_2$ is symplectic, it follows from the definition of ϕ, and the last statement in Theorem 8.8, that ϕ maps fibers of π_1 symplectomorphically onto fibers of π_2. To check the symplecticity of ϕ, it therefore suffices to check that the tangent map $\mathrm{T}\phi$ maps horizontal spaces of the symplectic connection on $\pi_1 : P_1 \to \mathcal{O}$ symplectically onto horizontal spaces of the symplectic connection on $\pi_2 : P_2 \to \mathcal{O}$. From the definition of ϕ and infinitesimal generators, it follows that

8.16 $$\mathrm{T}\phi \cdot \xi_{P_1}(g \cdot y) = \xi_{P_2}(\phi(g \cdot y)) \qquad (\xi \in \mathfrak{g},\ g \in G,\ y \in F) \ .$$

Since $\mathbf{J}^2(\phi(g \cdot y)) = \mathbf{J}^1(g \cdot y)$ (proven above), the symplecticity of $\mathrm{T}\phi$ on horizontal spaces follows immediately from 8.16 and Lemma 8.12. □

The π-invariant T-action

The remaining results of this chapter play a minor role in the sequel and may be skipped on a first reading. We call an action of a group H on P *π-invariant* if the fibers of π are H-invariant submanifolds. The action of T (as a subgroup of G) is not π-invariant, but it does leave the symplectic cross-section F invariant. We seek to extend this action on F to a π-invariant action. It is convenient to begin with a slightly more general result:

8.17 Lemma (Extension Lemma). *Let $(P, \omega, G, \mathbf{J})$ be a Hamiltonian G-space with regular momenta, $\pi : P \to \mathcal{O}$ the associated symplectic fibration, and $(F, \omega_F, T, \mathbf{J}^F)$ the symplectic cross-section. Suppose that an Abelian group H acts on F in a Hamiltonian fashion and let $\mathbf{K} : F \to \mathfrak{h}^*$ be an associated equivariant momentum map. Assume furthermore that:*

1. *The actions of H and T commute.*
2. *$\mathbf{K}$ is T-invariant.*
3. *$\mathbf{J}|F$ is H-invariant.*

Then, denoting the action of H on F by $(h, y) \mapsto h \odot y$, there exists a well-defined extension to P given by

$$h \odot (g \cdot y) \equiv g \cdot (h \odot y) \qquad (h \in H, g \in G, y \in F) \ .$$

The extended action of H leaves the fibers of π invariant and is Hamiltonian. An associated momentum map is the well-defined extension of $\mathbf{K}$ to a map $\mathbf{K} : P \to \mathfrak{h}^*$ given by

$$\mathbf{K}(g \cdot y) \equiv \mathbf{K}(y) \qquad (g \in G, y \in F) \ .$$

Moreover, the properties 1 and 2 above extend as follows:

4. *The extended action of H commutes with the action of G.*
5. *The extended momentum map $\mathbf{K} : P \to \mathfrak{h}^*$ is G-invariant.*
6. *$\mathbf{J}$ is invariant with respect to the extended action of H.*

The lemma, proven in Appendix C, follows from Theorem 8.8 and Lemma 8.12.

Taking $H \equiv T$, we obtain an extension of the action of T on F to a π-invariant action on P, with momentum map $\mathbf{K} : P \to \mathfrak{t}^*$ given by

$$\begin{aligned}\mathbf{K}(g \cdot y) \equiv \mathbf{J}^F(y) = \mathbf{J}(y)|\mathfrak{t} = i(\mathbf{J}(y)) &= i(\pi_{\mathcal{W}}(g \cdot \mathbf{J}(y))) \qquad \text{by 8.5} \\ &= (i \circ \pi_{\mathcal{W}} \circ \mathbf{J})(g \cdot y) \qquad \text{by the equivariance of } \mathbf{J} \\ &= \mathbf{j}(g \cdot y) \qquad (\mathbf{j} \text{ is defined in 8.7}) \ .\end{aligned}$$

Notice also that we can write $q \odot (g \cdot y) = gqg^{-1} \cdot (g \cdot y)$ $(q \in T,\ g \in G,\ y \in F)$. Summarizing:

8.18 Corollary. *If $(P, \omega, G, \mathbf{J})$ is a Hamiltonian G-space with regular momenta, and π the associated symplectic fibration, then there exists a well-defined π-invariant action $(q, x) \mapsto q \odot x$ of T on P given by*

$$q \odot (g \cdot y) \equiv gq \cdot y \qquad (q \in T, g \in G, y \in F) \ .$$

This action is Hamiltonian with momentum map $\mathbf{j} : P \to \mathfrak{t}_0^* \subset \mathfrak{t}^*$ defined by $\mathbf{j} \equiv i \circ \pi_{\mathcal{W}} \circ \mathbf{J}$. The actions of T and G_ν on a fiber F_ν $(\nu \in \mathcal{O})$ are conjugate in the following sense: Let $g \in G$ be such that $\nu = g \cdot \mu_0$ and define the isomorphism $\psi : T \to G_\nu$ by $\psi(q) \equiv gqg^{-1}$. (Recall that $T = G_{\mu_0}$, so that $gTg^{-1} = G_{g \cdot \mu_0} = G_\nu$.) Then $q \odot x = \psi(q) \cdot x$ for any $x \in F_\nu$.

CHAPTER 9

Action-Group Coordinates as a Symplectic Cross-Section

One can use symplectic cross-sections as a way of generating new symplectic manifolds[1]. In this chapter we show that the action-group model space $G \times \mathfrak{t}_0^*$ described in Part 1 can be realized as the symplectic cross-section of T^*G (with its 'irregular' points removed). We study the space $G \times \mathfrak{t}_0^*$ in some detail, supplying proofs of some facts stated in Part 1.

The symplectic cross-section of T^*G

Let Θ denote the *canonical one-form* on T^*G. This is defined by

9.1 $$\langle \Theta, \zeta \rangle = \langle \tau_{\mathrm{T}^*G}(\zeta), \mathrm{T}\tau_G^* \cdot \zeta \rangle \qquad (\zeta \in \mathrm{T}(\mathrm{T}^*G)) .$$

The maps $\tau_{\mathrm{T}^*G} : \mathrm{T}(\mathrm{T}^*G) \to \mathrm{T}^*G$ and $\tau_G^* : \mathrm{T}^*G \to G$ denote the canonical projections. A natural symplectic structure on T^*G is $\omega \equiv -d\Theta$.

A (left) action of G on itself is given by

$$g \cdot h \equiv R_{g^{-1}}(h) \qquad (g, h \in G) ,$$

where $R_g(h) \equiv hg$. Recall that the (covariant) *cotangent lift* $\mathrm{T}_*\phi : \mathrm{T}^*G \to \mathrm{T}^*G$ of a diffeomorphism $\phi : G \to G$ is defined by

$$\langle \mathrm{T}_*\phi \cdot \alpha, v \rangle \equiv \langle \alpha, \mathrm{T}\phi^{-1} \cdot v \rangle \qquad (\alpha \in \mathrm{T}_g^*G,\ v \in \mathrm{T}_{\phi(g)}G,\ g \in G) .$$

We make G act on T^*G by cotangent-lifting the above action of G on itself:

$$g \cdot x \equiv \mathrm{T}_*R_{g^{-1}} \cdot x \qquad (g \in G, x \in \mathrm{T}^*G) .$$

It follows from general results concerning cotangent-lifted actions (see, e.g., Marsden and Ratiu (1994)) that the above action of G on T^*G is Hamiltonian, with an equivariant momentum map $\mathbf{J} : \mathrm{T}^*G \to \mathfrak{g}^*$ given by

$$\mathbf{J}(x) \equiv \mathrm{T}_*L_{g^{-1}} \cdot x \qquad (x \in \mathrm{T}^*G,\ g \equiv \tau_G^*(x)) ,$$

where $L_g(h) \equiv gh$ $(g, h \in G)$. The open dense G-invariant subset $P \equiv \mathbf{J}^{-1}(\mathfrak{g}_{\mathrm{reg}}^*) \subset \mathrm{T}^*G$ is a Hamiltonian G-space with regular momenta. Each element of P is of the form $\mathrm{T}_*L_g \cdot \mu$ for some $g \in G$ and $\mu \in \mathfrak{g}_{\mathrm{reg}}^*$. (Indeed $(g, \mu) \mapsto \mathrm{T}_*L_g \cdot \mu : G \times \mathfrak{g}_{\mathrm{reg}}^* \to P$ is a diffeomorphism.) The symplectic fibration $\pi : P \to \mathcal{O}$ (see 8.7) is given by

$$\pi(\mathrm{T}_*L_g \cdot \mu) = \pi_{\mathcal{O}}(\mu) \qquad (g \in G,\ \mu \in \mathfrak{g}_{\mathrm{reg}}^*) .$$

Therefore, the symplectic cross-section of $(P, \omega, G, \mathbf{J})$ is given by

$$F \equiv \{\mathrm{T}_*L_g \cdot \mu \mid g \in G,\ \mu \in \mathcal{W}\} .$$

---

[1]This is certainly *not* our main reason for introducing the cross-section technology. The more significant application is Theorem 10.2 in the next chapter.

For each $\alpha \in \mathfrak{g}^*$, let α_G denote the left-invariant one-form on G with $\alpha_G(\mathrm{id}_G) = \alpha$ (here viewing one-forms on G as sections of $G \to \mathrm{T}^*G$). Then the embedding $\phi : G \times \mathfrak{t}_0^* \hookrightarrow \mathrm{T}^*G$ defined by $\phi(g,p) \equiv (i^{-1}(p))_G(g) = \mathrm{T}_*L_g \cdot i^{-1}(p)$ is a diffeomorphism onto F. (Since Theorem 8.8 says that $F \subset P$ is a symplectic submanifold, this proves Proposition 2.1, p. 10.) We define $\omega_G^* \equiv \phi^*\omega$, and call $(G \times \mathfrak{t}_0^*, \omega_G^*)$ the *action-group model space* of G.

In deriving an explicit formula for the symplectic structure ω_G^*, it will be convenient to have a concrete way of expressing vectors tangent to $G \times \mathfrak{t}_0^*$. To this end, define for each $(\xi, \tau) \in \mathfrak{g} \times \mathfrak{t}^*$ the vector field $(\xi,\tau)_{\mathrm{vf}}$ on $G \times \mathfrak{t}_0^*$ by

$$(\xi,\tau)_{\mathrm{vf}}(g,p) \equiv \frac{d}{dt}(g\exp(t\xi), p + t\tau)\Big|_{t=0} \quad . \tag{9.2}$$

Since $(\xi,\tau) \mapsto (\xi,\tau)_{\mathrm{vf}}(g,p) : \mathfrak{g} \times \mathfrak{t}^* \to \mathrm{T}_{(g,p)}(G \times \mathfrak{t}_0^*)$ is an isomorphism at every $(g,p) \in G \times \mathfrak{t}_0^*$, every vector tangent to $G \times \mathfrak{t}_0^*$ is uniquely expressible in the form $(\xi,\tau)_{\mathrm{vf}}(g,p)$. One verifies that[2]

$$[(\xi_1,\tau_1)_{\mathrm{vf}}, (\xi_2,\tau_2)_{\mathrm{vf}}] = ([\xi_1,\xi_2], 0)_{\mathrm{vf}} \quad . \tag{9.3}$$

9.4 Lemma. *Define $\Theta_G^* = \phi^*\Theta$, so that $\omega_G^* = -d\Theta_G^*$. Then*

$$\langle \Theta_G^*, (\xi,\tau)_{\mathrm{vf}}(g,p)\rangle = \langle p, \sigma(\xi)\rangle \quad ,$$

where $\sigma : \mathfrak{g} \to \mathfrak{t}$ is the projection onto $\mathfrak{t}$ along $\mathfrak{t}^\perp \equiv [\mathfrak{g},\mathfrak{t}]$. Furthermore,

$$\omega_G^*\big(\,(\xi_1,\tau_1)_{\mathrm{vf}}(g,p), (\xi_2,\tau_2)_{\mathrm{vf}}(g,p)\,\big) = \langle \tau_2, \sigma(\xi_1)\rangle - \langle \tau_1, \sigma(\xi_2)\rangle + \langle p, \sigma[\xi_1,\xi_2]\rangle \quad .$$

The above expression for Θ_G^* also appears in Dazord and Delzant (1987, Section 5) (who obtain it via a different route).

Proof. One computes using 9.1

$$\begin{aligned}
\langle \Theta_G^*, (\xi,\tau)_{\mathrm{vf}}(g,p)\rangle &= \langle \Theta, \mathrm{T}\phi \cdot (\xi,\tau)_{\mathrm{vf}}(g,p)\rangle \\
&= \Big\langle \mathrm{T}_*L_g \cdot i^{-1}(p), \frac{d}{dt}\tau_G^*\Big(\phi(g\exp(t\xi), p + t\tau)\Big)\Big|_{t=0}\Big\rangle \\
&= \Big\langle i^{-1}(p), \mathrm{T}L_{g^{-1}} \cdot \frac{d}{dt}\tau_G^*(\mathrm{T}_*L_{g\exp(t\xi)} \cdot i^{-1}(p + t\tau))\Big|_{t=0}\Big\rangle \\
&= \Big\langle i^{-1}(p), \frac{d}{dt}g^{-1}g\exp(t\xi)\Big|_{t=0}\Big\rangle \\
&= \langle i^{-1}(p), \xi\rangle = \langle p, \sigma(\xi)\rangle \quad .
\end{aligned} \tag{9.5}$$

From this we compute

$$\big(\,(\xi_1,\tau_1)_{\mathrm{vf}} \lrcorner\, d((\xi_2,\tau_2)_{\mathrm{vf}} \lrcorner\, \Theta_G^*)\,\big)(g,p) = \frac{d}{dt}\langle p + t\tau_1, \sigma(\xi_2)\rangle\Big|_{t=0} = \langle \tau_1, \sigma(\xi_2)\rangle \quad . \tag{9.6}$$

Applying the well-known identity

$$v \lrcorner\, u \lrcorner\, d\beta = u \lrcorner\, d(v \lrcorner\, \beta) - v \lrcorner\, d(u \lrcorner\, \beta) - [u,v] \lrcorner\, \beta \quad ,$$

with $\beta = \Theta_G^*$, we obtain from 9.3, 9.5 and 9.6 the formula for ω_G^* stated in the lemma. □

[2] We identify the Lie algebra of a Lie group with the *left* invariant vector fields, equipped with Jacobi-Lie bracket; our sign convention for the latter is the same as Abraham, Marsden and Ratiu (1988, §4.2.20).

9.7 Remark. If G is a torus $\mathbb{T}^k$, then $G \times \mathfrak{t}_0^* \cong \mathbb{T}^k \times \mathbb{R}^k$ and ω_G^* is identifiable with the canonical symplectic structure $\sum_j dq_j \wedge dp_j$. In this sense $G \times \mathfrak{t}_0^*$ (G an arbitrary compact connected Lie group) is a non-Abelian generalization of action-angle coordinates.

Action-group coordinates as a Hamiltonian G-space

Since we have realized $G \times \mathfrak{t}_0^*$ as a symplectic cross-section, the maximal torus T acts on $G \times \mathfrak{t}_0^*$. More significant for us, however, is the observation that G acts on $G \times \mathfrak{t}_0^*$:

9.8 Lemma. *The action of G on $G\times\mathfrak{t}_0^*$ defined by $g\cdot(h,p) \equiv (gh,p)$ is Hamiltonian, with equivariant momentum map $\mathbf{J}^G : G \times \mathfrak{t}_0^* \to \mathfrak{g}^*$ given by $\mathbf{J}^G(g,p) \equiv g \cdot i^{-1}(p) = \mathrm{Ad}^*_{g^{-1}}\, i^{-1}(p)$. This action makes $(G \times U, \omega_G^*, G, \mathbf{J}^G)$ a Hamiltonian G-space with regular momenta for any open set $U \subset \mathfrak{t}_0^*$.*

PROOF. Imitate the construction in Example 3.2, p. 16. □

9.9 Remark. The action of T on the Hamiltonian G-space $G \times \mathfrak{t}_0^*$ given in 8.18 is defined by $q \odot (g,p) \equiv (gq,p)$. According to 8.18, this action is Hamiltonian, with a momentum map $\mathbf{j}^G : G \times \mathfrak{t}_0^* \to \mathfrak{t}_0^*$ given by $\mathbf{j}^G(g,p) = (i \circ \pi_{\mathcal{W}} \circ \mathbf{J}^G)(g,p) = p$.

The following proposition, whose easy proof is left to the reader, shows that the symplectic cross-section of the action-group model space of G is the action-group model space of T, i.e., conventional action-*angle* coordinates (by Remark 9.7).

9.10 Proposition.
The symplectic cross-section of $(G \times U, \omega_G^*, G, \mathbf{J}^G)$ is $(T \times U, \omega_T^*, T, \mathbf{J}^T)$.

Hamiltonian vector fields in action-group coordinates

In our applications to perturbation theory described in Part 1, it was essential to have a concrete way of writing down the equations of motion in action-group coordinates. These equations were derived from an expression for Hamiltonian vector fields that was stated without proof. We now derive that expression explicitly. (In Part 1 it was convenient to identify $G\times\mathfrak{t}_0^*$ with $G\times\mathfrak{t}_0$ using some Ad-invariant inner product on $\mathfrak{g}$. Here we work throughout with $G \times \mathfrak{t}_0^*$ and leave it to the interested reader to translate our results into the form used in Part 1.)

9.11 Lemma. *Write $\mathfrak{t}^\perp \equiv [\mathfrak{g},\mathfrak{t}]$ and $\underline{\mathfrak{t}}^\perp \equiv \mathrm{Ann}\,\mathfrak{t}$ (notation as in Chap. 1 of Part 1). Then for all $p \in \mathfrak{t}_0^*$ the map*

$$\xi \mapsto \mathrm{ad}^*_\xi(i^{-1}(p)) : \mathfrak{t}^\perp \to \underline{\mathfrak{t}}^\perp$$

is an isomorphism.

PROOF. Since $\dim \mathfrak{t}^\perp = \dim \underline{\mathfrak{t}}^\perp$, it suffices to show that $\mathrm{ad}^*_\xi\, i^{-1}(p) = 0$ implies $\xi = 0$. Write $\mu \equiv i^{-1}(p) \in \mathcal{W} \subset \underline{\mathfrak{t}} \cap \mathfrak{g}^*_{\mathrm{reg}}$, so that $\mathfrak{t} = \mathfrak{g}_\mu$ (by, e.g., Corollary 1.6.2, p. 9). Supposing that $\xi \in \mathfrak{t}^\perp$ satisfies $\mathrm{ad}^*_\xi\, i^{-1}(p) = 0$, we have $\mathrm{ad}^*_\xi\, \mu = 0$, i.e., $\xi \in \mathfrak{g}_\mu = \mathfrak{t}$. But $\mathfrak{t} \cap \mathfrak{t}^\perp = \{0\}$ (by, e.g., Theorem 1.2.8, p. 8), so that $\xi = 0$. □

For any $p \in \mathfrak{t}_0^*$, we define $\lambda_p : \underline{\mathfrak{t}}^\perp \to \mathfrak{t}^\perp$ to be the inverse of the map described in 9.11 above. For future reference, let us record that this means

$$\lambda_p\left(\mathrm{ad}^*_\xi\, i^{-1}(p)\right) = \xi \qquad (\xi \in \mathfrak{t}^\perp,\ p \in \mathfrak{t}_0^*)\ . \tag{9.12}$$

For any smooth function $f : G \times \mathfrak{t}_0^* \to \mathbb{R}$ define the vector-valued functions $\frac{\partial f}{\partial g} : G \times \mathfrak{t}_0^* \to \mathfrak{g}^*$ and $\frac{\partial f}{\partial p} : G \times \mathfrak{t}_0^* \to \mathfrak{t}$ by

$$\langle \frac{\partial f}{\partial g}(g,p), \xi \rangle \equiv \langle df, (\xi, 0)_{\mathrm{vf}}(g,p) \rangle = \frac{d}{dt} f(g \exp(t\xi), p)\Big|_{t=0} \qquad (\xi \in \mathfrak{g}) \ ,$$

$$\langle \tau, \frac{\partial f}{\partial p}(g,p) \rangle \equiv \langle df, (0, \tau)_{\mathrm{vf}}(g,p) \rangle = \frac{d}{dt} f(g, p + t\tau)\Big|_{t=0} \qquad (\tau \in \mathfrak{t}^*) \ .$$

9.13 Proposition (Hamiltonian vector fields on $G \times \mathfrak{t}_0^*$).
The Hamiltonian vector field X_f corresponding to a function f is given by

$$X_f(g,p) = (\xi, \tau)_{\mathrm{vf}}(g,p) \ ,$$

where

$$\xi \equiv \frac{\partial f}{\partial p}(g,p) + \Lambda_p \frac{\partial f}{\partial g}(g,p)$$

$$\tau \equiv -i\sigma^* \frac{\partial f}{\partial g}(g,p)$$

$$\Lambda_p \equiv \lambda_p \circ (\mathrm{id} - \sigma^*) \ .$$

Here $\sigma^* : \mathfrak{g}^* \to \underline{\mathfrak{t}}$ denotes the projection along $\underline{\mathfrak{t}}^\perp \equiv \mathrm{Ann}\, \mathfrak{t}$, i.e., $\langle \sigma^* \mu, \xi \rangle = \langle \mu, \sigma\xi \rangle$ $(\mu \in \mathfrak{g}^*, \xi \in \mathfrak{g})$.

Proof. Define $\xi \in \mathfrak{g}$ and $\tau \in \mathfrak{t}^*$ as in the statement of the lemma. It suffices to verify that the one-forms $(\xi, \tau)_{\mathrm{vf}} \lrcorner\, \omega_G^*$ and df agree on elements of $\mathrm{T}_{(g,p)}(G \times \mathfrak{t}_0^*)$. Let $x \in \mathfrak{g}$ and $y \in \mathfrak{t}^*$ be given. Then, by Lemma 9.4,

$$\begin{aligned} &\langle (\xi,\tau)_{\mathrm{vf}} \lrcorner\, \omega_G^*, (x,y)_{\mathrm{vf}}(g,p) \rangle = \langle y, \sigma\xi \rangle - \langle \tau, \sigma x \rangle + \langle p, \sigma[\xi, x] \rangle \\ &= \langle i\sigma^* \frac{\partial f}{\partial g}(g,p), \sigma x \rangle + \langle y, \frac{\partial f}{\partial p}(g,p) \rangle + \langle i^{-1}(p), [\Lambda_p \frac{\partial f}{\partial g}(g,p), x] + [\frac{\partial f}{\partial p}(g,p), x] \rangle \end{aligned}$$

9.14

$$= \langle \frac{\partial f}{\partial g}(g,p), \sigma x \rangle + \langle y, \frac{\partial f}{\partial p}(g,p) \rangle + \langle i^{-1}(p), \Lambda_p \frac{\partial f}{\partial g}(g,p), x] \rangle \ .$$

We have used the fact $\langle i^{-1}(p), [\frac{\partial f}{\partial p}(g,p), x] \rangle = 0$, which is true because $i^{-1}(p) \in \mathrm{Ann}[\mathfrak{t}, \mathfrak{g}]$ and $\frac{\partial f}{\partial p}(g,p) \in \mathfrak{t}$. Noting that $\Lambda_p \frac{\partial f}{\partial g}(g,p) \in \mathfrak{t}^\perp$ and using 9.12,

$$\begin{aligned} \langle i^{-1}(p), [\Lambda_p \frac{\partial f}{\partial g}(g,p), x] \rangle &= \langle \lambda_p^{-1} \Lambda_p \frac{\partial f}{\partial g}(g,p), x \rangle \\ &= \langle (\mathrm{id} - \sigma^*) \frac{\partial f}{\partial g}(g,p), x \rangle \\ &= \langle \frac{\partial f}{\partial g}(g,p), (\mathrm{id} - \sigma) x \rangle \ , \end{aligned}$$

so that 9.14 becomes

$$\begin{aligned} \langle (\xi,\tau)_{\mathrm{vf}} \lrcorner\, \omega_G^*, (x,y)_{\mathrm{vf}}(g,p) \rangle &= \langle \frac{\partial f}{\partial g}(g,p), x \rangle + \langle y, \frac{\partial f}{\partial p}(g,p) \rangle \\ &= \langle df, (x,y)_{\mathrm{vf}}(g,p) \rangle \ . \end{aligned}$$

Since $x \in \mathfrak{g}$ and $y \in \mathfrak{t}^*$ were arbitrary, this proves our assertion. □

The Poisson bracket in action-group coordinates

The Poisson bracket on $G \times \mathfrak{t}_0^*$ is defined by $\{f,h\} \equiv X_h \lrcorner X_f \lrcorner \omega_G^*$.

9.15 Corollary (The Poisson bracket on $G \times \mathfrak{t}_0^*$).
Dropping the '(g,p)' argument from '$\frac{\partial f}{\partial g}(g,p)$,' etc., the Poisson bracket on $G \times \mathfrak{t}_0^*$ is given by

$$\{f,h\}(g,p) = \langle \frac{\partial f}{\partial g}, \frac{\partial h}{\partial p} \rangle - \langle \frac{\partial h}{\partial g}, \frac{\partial f}{\partial p} \rangle - \langle p, \sigma[\Lambda_p \frac{\partial f}{\partial g}, \Lambda_p \frac{\partial h}{\partial g}] \rangle,$$

where $\Lambda_p : \mathfrak{g}^* \to \mathfrak{g}$ is the map defined in Lemma 9.13 above.

PROOF. Imitate the proof of Lemma 2.14, p. 13. □

CHAPTER 10

Constructing Action-Group Coordinates

Noticing that equivalent Hamiltonian G-spaces (see 8.1) have identical momentum map images, we introduce the following terminology:

10.1 Definition. Let $(P, \omega, G, \mathbf{J})$ be a Hamiltonian G-space and define $V \equiv \mathbf{J}(P)$ and $U \equiv i(V \cap \mathcal{W})$. (If G is Abelian, then $U = V$.) We say that P *admits G-compatible action-group coordinates* if $(P, \omega, G, \mathbf{J})$ is equivalent to $(G \times U, \omega_G, G, \mathbf{J}^G)$. We sometimes distinguish the Abelian case (G, a torus) by saying that P *admits G-compatible action-**angle** coordinates.*

The non-Abelian case

Here is our main result. It follows immediately from 8.14 and 9.10.

10.2 Theorem.
A Hamiltonian G-space $(P, \omega, G, \mathbf{J})$ with regular momenta admits G-compatible action-group coordinates $\Leftrightarrow$ its symplectic cross-section $(F, \omega_F, T, \mathbf{J}^F)$ admits T-compatible action-angle coordinates. Indeed if U denotes the image of $\mathbf{J}^F$, and $\varphi : T \times U \xrightarrow{\sim} F$ an equivalence, then a well-defined equivalence $\phi : G \times U \xrightarrow{\sim} P$ is given by

$$\phi(g \cdot y) \equiv g \cdot \varphi(y) \qquad (g \in G,\ y \in F) .$$

The existence of action-angle coordinates can be characterized as follows.

10.3 Proposition. *Let T be a torus and $(F, \omega, T, \mathbf{J})$ a Hamiltonian T-space. Then F admits T-compatible action-angle coordinates $\Leftrightarrow$ the following conditions hold:*

1. *T acts freely and the space is geometrically integrable (see 3.7, p. 18).*
2. *$\mathbf{J} : F \to U \equiv \mathbf{J}(F)$ has connected fibers.*
3. *$\mathbf{J} : F \to U$ admits a (global) isotropic section $s : U \to F$.*

Recall that a section $s : U \to F$ is *isotropic* if $s^*\omega = 0$. In this Abelian case geometric integrability is equivalent to the condition $\dim F = 2 \dim U$. In that case the image of a section $s : U \to F$ has half the dimension of the ambient space. A section whose image has this dimension and is isotropic is usually called a *Lagrangian* section.

Proof of $\Rightarrow$. That conditions 1 and 2 hold is obvious. Condition 3 follows since the map $p \mapsto (q, p) : U \to T \times U$ is an isotropic section (for any $q \in T$). $\square$

Proof of $\Leftarrow$. Since T acts freely, $\mathbf{J} : F \to U$ is a surjective submersion (apply identity 8.3). By the pre-image theorem the fibres of $\mathbf{J}$ are submanifolds of F of dimension $\dim F - \dim T$. Since the space is geometrically integrable this dimension is $\dim T$. By momentum map equivariance — which amounts to *invariance* since T is Abelian — the connected components of a fibre are therefore T-orbits. It follows from condition 2 that each fibre is a *single* T-orbit. Consequently, as T acts freely, the map $\phi : T \times U \to F$ defined by $\phi(q,p) \equiv q \cdot s(p)$ is a diffeomorphism.

Now ϕ is clearly T-equivariant, and we have $(\mathbf{J} \circ \phi)(q,p) = p = \mathbf{J}^T(q,p)$, since $\mathbf{J}$ is T-invariant and s is a section. So to show that ϕ is an equivalence, it remains to show that ϕ is symplectic. This means showing that $\phi^*\omega = \omega_T^*$ where, according to Lemma 9.4, ω_T^* is given by

10.4
$$\omega_T^*\left((\xi_1,\tau_1)_{\mathrm{vf}}(q,p), (\xi_2,\tau_2)_{\mathrm{vf}}(q,p) \right) = \langle \tau_2, \xi_1 \rangle - \langle \tau_1, \xi_2 \rangle$$
$$\left((q,p) \in T \times U;\ \xi_1, \xi_2 \in \mathfrak{t};\ \tau_1, \tau_2 \in \mathfrak{t}^* \right) \ .$$

We will compare $\phi^*\omega$ and ω_T^* on pairs of vectors of the form $(\xi,\tau)_{\mathrm{vf}}(q,p)$. First note that by a straightforward computation, we have

$$\mathrm{T}\phi \cdot (\xi,\tau)_{\mathrm{vf}}(q,p) = \xi_F(\phi(q,p)) + (\mathrm{T}\Phi_q \circ \mathrm{T}s) \cdot \frac{d}{dt}(p + t\tau)\Big|_{t=0} \ ,$$

where the diffeomorphism $\Phi_q : F \to F$ is defined by $\Phi_q(x) \equiv q \cdot x$. Using this we compute

10.5
$$\begin{aligned}\phi^*\omega(\,(\xi_1,\tau_1)_{\mathrm{vf}}(q,p), (\xi_2,\tau_2)_{\mathrm{vf}}(q,p)\,) = &\\ \omega(\,(\xi_1)_F(\phi(q,p)), (\xi_2)_F(\phi(q,p))\,)&\\ + \omega\left((\xi_1)_F(\phi(q,p)), (\mathrm{T}\Phi_q \circ \mathrm{T}s) \cdot \frac{d}{dt}(p + t\tau_2)\Big|_{t=0} \right)&\\ - \omega\left((\xi_2)_F(\phi(q,p)), (\mathrm{T}\Phi_q \circ \mathrm{T}s) \cdot \frac{d}{dt}(p + t\tau_1)\Big|_{t=0} \right)&\\ + \omega\left((\mathrm{T}\Phi_q \circ \mathrm{T}s) \cdot \frac{d}{dt}(p + t\tau_1)\Big|_{t=0}, (\mathrm{T}\Phi_q \circ \mathrm{T}s) \cdot \frac{d}{dt}(p + t\tau_2)\Big|_{t=0} \right)& \ .\end{aligned}$$

The first term on the right-hand side vanishes by 8.12.1, since T is Abelian. The last term vanishes because Φ_q is symplectic and $s : U \to F$ is isotropic. The remaining terms are of the form

$$\begin{aligned}&\omega\left(\xi_F(\phi(q,p)), (\mathrm{T}\Phi_q \circ \mathrm{T}s) \cdot \frac{d}{dt}(p + t\tau)\Big|_{t=0} \right)\\ &= \omega\left(\xi_F(q,p), \mathrm{T}s \cdot \frac{d}{dt}(p + ts)\Big|_{t=0} \right) \quad \text{since } \Phi_q \text{ is symplectic and } \Phi_q^*\xi_F = \xi_F\\ &= \langle dJ_\xi, \mathrm{T}s \cdot \frac{d}{dt}(p + t\tau) \rangle \qquad \text{since } \xi_F = X_{J_\xi}\\ &= \langle \frac{d}{dt}(\mathbf{J}(s(p + t\tau)), \xi \rangle = \langle \tau, \xi \rangle \qquad \text{since } \mathbf{J} \circ s = \mathrm{id} \ (s \text{ is a section}).\end{aligned}$$

From this we see that 10.5 and 10.4 have the same right hand sides, so that $\phi^*\omega = \omega_T^*$ as claimed.

□

Of course our main interest in Proposition 10.3 is the case in which F is the symplectic cross-section of some Hamiltonian G-space P (G non-Abelian). In that case, some of the conditions listed in 10.3 can be cast as conditions on the space P. This will be useful later on.

10.6 Lemma.

Let $(P, \omega, G, \mathbf{J})$ be a Hamiltonian G-space with regular momenta and $(F, \omega_F, T, \mathbf{J}^F)$ its symplectic cross-section. Then

1. *T acts freely and $(F, \omega_F, T, \mathbf{J}^F)$ is geometrically integrable*
   $\Leftrightarrow$ G acts freely and $(P, \omega, G, \mathbf{J})$ is geometrically integrable (see 3.7, p. 18).
2. *Every fiber of $\mathbf{J}^F : F \to U \equiv \mathbf{J}^F(F)$ is homeomorphic to a fiber of $\mathbf{J} : P \to V \equiv \mathbf{J}(P)$ and conversely. (In particular, $\mathbf{J}^F$ has connected fibers $\Leftrightarrow$ $\mathbf{J}$ has connected fibers.)*

PROOF. That $T \subset G$ acts freely on $F \subset P$ when G acts freely on P is trivial. Conversely, suppose T acts freely on F, and let $g \in G$ be such that $g \cdot x = x$ for some $x \in P$. Then by the equivariance of the symplectic fibration $\pi : P \to \mathcal{O}$, we have $g \in G_\nu$, where $\nu \equiv \pi(x)$. Since G acts transitively on $\mathcal{O}$, there is $h \in G$ such that $\nu = h \cdot \mu_0$. Then $G_\nu = G_{h \cdot \mu_0} = hG_{\mu_0}h^{-1} = hTh^{-1}$, so that $g = hqh^{-1}$ for some $q \in T$. Then $g \cdot x = x$ implies $q \cdot (h^{-1} \cdot x) = h^{-1} \cdot x$. Since $h^{-1} \cdot x \in F_{\mu_0} = F$ (again by the equivariance of π), and we are supposing that T acts freely on F, it follows that $q = \mathrm{id}$, so that $g = hh^{-1} = \mathrm{id}$. This shows that G acts freely. Since $\dim P = \dim \mathcal{O} + \dim F = (\dim G - \operatorname{rank} G) + \dim F$, and $\operatorname{rank} G = \dim T$, we have

$$\dim F = 2 \dim T \quad \Leftrightarrow \quad \dim P = \dim G + \operatorname{rank} G \ .$$

This shows that F is geometrically integrable if and only if P is geometrically integrable. So 1 is true.

Each fiber of $\mathbf{J}$ lies entirely in a fiber of the symplectic fibration $\pi : P \to \mathcal{O}$, and is therefore (by, e.g., 8.8) homeomorphic to a fiber of $\mathbf{J}$ lying entirely in the fiber $F = F_{\mu_0}$. Now by definition $\mathbf{J}^F = i(\mathbf{J}|F)$, where $i : \underline{\mathfrak{t}} \to \mathfrak{t}^*$ is an isomorphism. So every fiber of $\mathbf{J}^F$ is literally a fiber of $\mathbf{J}$, and every fiber of $\mathbf{J}$ is homeomorphic to a fiber of $\mathbf{J}^F$. So 2 is true. □

The following observation is useful in constructing explicit action-group coordinates in concrete examples:

10.7 Proposition. *Let $(P, \omega, G, \mathbf{J})$ be a Hamiltonian G-space with regular momenta and $(F, \omega_F, T, \mathbf{J}^F)$ its symplectic cross-section. Assume that F admits T-compatible action-angle coordinates (so that P admits G-compatible action-group coordinates, by Theorem 10.2) and let $s : U \to F$ be a Lagrangian section (whose existence is guaranteed by Proposition 10.3). Then a realization of G-compatible action-group coordinates is given by the equivalence $\phi : G \times U \xrightarrow{\sim} P$ defined explicitly by $\phi(g, p) \equiv g \cdot s(p)$.*

PROOF. Let $s : U \to F$ be a Lagrangian section. As in the proof of Proposition 10.3, we have an equivalence $\varphi : T \times U \xrightarrow{\sim} F$ given by $\varphi(q, p) \equiv q \cdot s(p)$. Applying Theorem 10.2, a well-defined equivalence $\phi : G \times U \to P$ is given by

$$\phi(g \cdot (q, p)) \equiv g \cdot \varphi(q, p) \qquad (g \in G,\ q \in T \subset G,\ p \in U) \ .$$

But this means $\phi(gq, p) = (gq) \cdot s(p)$ for any $g \in G$, $q \in T$, $p \in U$. In particular $\phi(g, p) = g \cdot s(p)$, as claimed. □

10.8 Example (Geodesic motions on S^2). In the example of geodesic motions on S^2 discussed in Chap. 8, the map $\varphi : S^1 \times (0, \infty) \to F$ defines an equivalence between the space

$$(T \times U, \omega_T, T, \mathbf{J}^T) = (S^1 \times (0, \infty), d\theta \wedge dI, S^1, I)$$

and the cross-section $(F, \omega_F, T, \mathbf{J}^F)$. In other words, F admits T-compatible action-angle coordinates. The map $I \mapsto (\theta, I) : (0, \infty) \to S^1 \times (0, \infty)$ is a Lagrangian section for any θ, so that a Lagrangian section $s : (0, \infty) \to F$ is given by

$$s(I) \equiv \varphi(0, I) = (e_1, Ie_2) \ .$$

The action-group model space for G is (up to the obvious identifications) $\mathrm{SO}(3) \times (0, \infty)$ (see, e.g., Example 3.4, p. 17). Applying Proposition 10.7, a realization of G-compatible action-group coordinates for P is given by the equivalence $\phi : \mathrm{SO}(3) \times (0, \infty) \to P$ defined by

$$\phi(g, p) \equiv g \cdot s(p) = (ge_1, pge_2) \ .$$

The SO(3)-invariant Hamiltonian $H(q, v) = \frac{1}{2} \|v\|^2$ is pulled back to $H \circ \phi(g, p) = h(p) \equiv \frac{1}{2}p^2$.

The Abelian case

Theorem 10.2 reduces the construction of action-group coordinates to that of action-angle coordinates. The remainder of this chapter revisits in some detail the classical problem of constructing action-angle coordinates, from the present Hamiltonian G-space point of view. (The necessary and sufficient conditions given in Proposition 10.3 are not always the most practical.)

Suppose then that $(F, \omega, T, \mathbf{J})$ is a Hamiltonian T-space on which a torus T is acting freely, and assume that the space is geometrically integrable. These assumptions are easy to check. Then, as we observed in the proof of Proposition 10.3, $\mathbf{J} : F \to U \equiv \mathbf{J}(F)$ is a surjective submersion, and the connected components of the fibers of $\mathbf{J}$ are T-orbits. Now unless one assumes that the fibers are *compact*, serious technical difficulties impede further progress[1]. Similar difficulties arise in attempts to generalize momentum map 'convexity theorems' to the non-compact case (see, e.g., Hilgert et al. (1994)[2]).

If the fibers of $\mathbf{J}$ are compact, or equivalently, have a finite number of connected components[3], then $\mathbf{J} : F \to U$ is a locally trivial fiber bundle; see, e.g. Sharpe (1997, 2.8.3(A) & 2.7.8). We consider two scenarios:

Scenario 1. Suppose that $\mathbf{J} : F \to U$ has connected components. (Checking this is not always trivial.) Then $\mathbf{J} : F \to U$ is a Lagrangian fibration with connected fibers (which are also compact, by assumption), to which the study of Duistermaat (1980) is applicable. That the fibers of $\mathbf{J}$ are Lagrangian follows from 8.12.2 and our integrability assumption. Duistermaat characterizes the obstruction to the existence of a (not necessarily Lagrangian) section $s : U \to F$ in terms of the Chern class ν of $\mathbf{J} : F \to U$. A section exists if and only if $\nu = 0$. In that case the de Rham cohomology class $[s^*\omega] \in H^2(U)$ is the same *for all sections* s, and a *Lagrangian* section exists if and only if this class vanishes. A sufficient condition is that ω is exact.

Scenario 2. In this case we weaken the connectedness hypothesis but enforce a topological assumption on the momentum map image. We formulate our conclusions in the form of a theorem, whose proof is constructive, up to the existence of a Riemannian metric.

---

[1]Unless one works 'locally'; see 3.11, p. 19.

[2]Note that these results do not apply immediately here since we cannot assume that $\mathbf{J}$ is proper as a map *into* $\mathfrak{t}^*$.

[3]Remember that these fibers are *regular* submanifolds, by the preimage theorem, and that each component of a given fibre is a (compact) T-orbit.

10.9 Theorem. *Let $(F, \omega, T, \mathbf{J})$ be a Hamiltonian T-space on which a torus T is acting freely, and assume this space is geometrically integrable (i.e., $\dim F = 2\dim T$). Assume F is connected and that each fiber of $\mathbf{J} : F \to U \equiv \mathbf{J}(F)$ is compact, or has a finite number of connected components. Assume that U is smoothly contractible. Then each fiber is in fact connected, and there exists a Lagrangian section $s : U \to F$ of $\mathbf{J} : F \to U$. In particular, F admits T-compatible action-angle coordinates.*

Note that in examples it is often easy to check the compactness of the fibers of $\mathbf{J}$. The Euler-Poinsot rigid body (see Chap. 11) is a case in point.

PROOF. The last statement follows from the preceding ones and Proposition 10.3. We begin by showing that the fibers of $\mathbf{J}$ are connected. By the hypotheses and our earlier remarks, $\mathbf{J} : F \to U$ is a locally trivial fiber bundle, each fiber being a finite number of T-orbits. Proceed by constructing a covering space $\tilde{\mathbf{J}} : \tilde{U} \to U$ as follows: For each $p \in U$ let Q_p denote the set of connected components of $\mathbf{J}^{-1}(p)$, and for each $x \in F$ let $[x]$ denote the element of $Q_{\mathbf{J}(x)}$ that contains x. Define $\tilde{U} \equiv \cup_{p \in U} Q_p$ and $\tilde{\mathbf{J}} : \tilde{U} \to U$ by $\tilde{\mathbf{J}}([x]) \equiv \mathbf{J}(x)$. Using the local triviality of $\mathbf{J} : F \to U$ and the connectedness of F, it is not difficult to show that $\tilde{U}$ admits the structure of a smooth *connected* manifold with respect to which $\tilde{\mathbf{J}} : \tilde{U} \to U$ is a (smooth) covering map. The multiplicity (number of sheets) of this cover is the number of connected components in each fibre $\mathbf{J}^{-1}(p)$. But since $U \subset \mathfrak{t}_0^*$ is contractible, it is simply connected. From covering space theory it follows that $\tilde{\mathbf{J}} : \tilde{U} \to U$ is a diffeomorphism. The multiplicity of the cover is therefore one, proving that the fibres of $\mathbf{J}$ are connected.

The next step is to construct a section for $\mathbf{J} : F \to U$. Equip the bundle $\mathbf{J} : F \to U$ with a connection (i.e., a distribution of horizontal spaces). This is always possible by giving F the structure of a (smooth) Riemannian manifold. That one can do so in the C^∞ category is a standard partition of unity argument. In the real-analytic category this follows from the Whitney-Morrey-Grauert Embedding Theorem (Grauert, 1952). Now U contracts to some point $p_0 \in U$. That is, we have a smooth map $(p, t) \mapsto K_t(p) : U \times [0, 1] \to U$ with $K_0 = c_{p_0}$ and $K_1 = \mathrm{id}_U$, where $c_{p_0} : U \to U$ is the constant map $c_{p_0}(p) \equiv p_0$. Fix any $x_0 \in \mathbf{J}^{-1}(p_0)$ and define a smooth section $s : U \to F$ as follows: For any point $p \in U$ define a path $\gamma : [0, 1] \to U$ from p_0 to p by $\gamma(t) \equiv K_t(p)$. Use the connection to lift γ to a path $\tilde{\gamma} : [0, 1] \to F$ with $\tilde{\gamma}(0) = x_0$, and define $s(p) \equiv \tilde{\gamma}(1)$.

Since the fibers of $\mathbf{J}$ are precisely the T-orbits, and T acts freely, the smooth map $\phi : T \times U \to F$ defined by $\phi(q, p) \equiv q \cdot s(p)$ is a diffeomorphism.

Our final step will be to exhibit a *Lagrangian* section $\tilde{s} : U \to F$. This section will be constructed by modifying the existing section s using Moser's (1965) well-known 'Lie transforms' technique. To this end, we first show that ω is *exact*. Indeed, as $\mathbf{J} : F \to U$ is globally trivial and U is contractible, there exists a smooth deformation retract of F onto some fiber T'. The fibres are just T-orbits, which are Lagrangian since T is Abelian (apply the formula in 8.12.2). Because of the deformation retraction, the inclusion $\iota : T' \to F$ induces an isomorphism in de Rham cohomology. Since T' is Lagrangian $\iota^* \omega = 0$, implying that the cohomology class of ω vanishes, i.e., $\omega = d\Theta$ for some one-form Θ.

We are now ready to construct the Lagrangian section. Suppose α is a one-form on U. Then at each point $p \in U \subset \mathfrak{t}^*$ one may associate to α an element $\hat{\alpha}(p) \in \mathfrak{t}$

defined through

$$\langle \nu, \hat{\alpha}(p) \rangle = \langle \alpha, \nu \rangle \qquad (\nu \in \mathrm{T}_p U \cong \mathfrak{t}^*) \ .$$

One may next associate to α a vector field X^α on F, defined by

$$X^\alpha(x) \equiv (\hat{\alpha}(p))_F(x) \qquad (x \in F,\, p \equiv \mathbf{J}(x)) \ ,$$

where η_F denotes the infinitesimal generator of the T action associated with an element $\eta \in \mathfrak{t}$. In particular, by the definition of a momentum map

$$\text{10.10} \qquad \langle X^\alpha \lrcorner\, \omega, v \rangle = \langle dJ_{\hat{\alpha}(p)}, v \rangle \qquad (v \in \mathrm{T}_x F,\, x \in F,\, p \equiv \mathbf{J}(x)) \ .$$

The vector field X^α is tangent to the T-orbits which, as shown above, are precisely the fibers of $\mathbf{J}$. These fibers are compact, so that we have a well-defined flow associated with X^α, which we denote by $(x,t) \mapsto \alpha_t(x) : F \times (-\infty, \infty) \to F$. The time-one map α_1 is a diffeomorphism of F preserving the fibers of $\mathbf{J} : F \to U$. In general α_1 is *not* symplectic. The strategy is to define $\tilde{s} \equiv \alpha_1 \circ s$ and determine how α might be chosen such that $\tilde{s}^*\omega = 0$. We have

$$\begin{aligned} \tilde{s}^*\omega = s^*\alpha_1^*\omega &= s^*\omega + s^* \int_0^1 (\frac{d}{dt}\alpha_t^*\omega)dt \\ &= s^*\omega + s^* \int_0^1 (\alpha_t^*(X^\alpha \lrcorner\, d\omega) + \alpha_t^* d(X^\alpha \lrcorner\, \omega))dt \end{aligned}$$

$$\text{10.11} \qquad = s^*\omega + \int_0^1 s^*\alpha_t^* d(X^\alpha \lrcorner\, \omega)dt = s^*\omega + \int_0^1 d(s^*\alpha_t^*(X^\alpha \lrcorner\, \omega))dt \ .$$

For any $u \in \mathrm{T}_p U \cong \mathfrak{t}^*$ we have

$$\begin{aligned} \langle s^*\alpha_t^*(X^\alpha \lrcorner\, \omega), u \rangle &= \langle X^\alpha \lrcorner\, \omega, \mathrm{T}(\alpha_t \circ s) \cdot u \rangle \\ &= \langle dJ_{\hat{\alpha}(p)}, \mathrm{T}(\alpha_t \circ s) \cdot u \rangle \qquad \text{by 10.10} \\ &= \frac{d}{d\tau} J_{\hat{\alpha}(p)}((\alpha_t \circ s)(p + \tau u)) \Big|_{\tau=0} \\ &= \frac{d}{d\tau} \langle (\mathbf{J} \circ \alpha_t \circ s)(p + \tau u), \hat{\alpha}(p) \rangle \Big|_{\tau=0} \\ &= \langle \alpha, \frac{d}{d\tau}(p + \tau u)\Big|_{\tau=0} \rangle = \langle \alpha, u \rangle \ . \end{aligned}$$

Therefore $s^*\alpha_t^*(X^\alpha \lrcorner\, \omega) = \alpha$. Using this in 10.11 gives $\tilde{s}^*\omega = s^*\omega + d\alpha$. Choosing $\alpha = s^*\Theta$ gives $\tilde{s}^*\omega = 0$, so that $\tilde{s}$ is a Lagrangian section. □

The following corollary was stated without proof as Theorem 3.10, p. 19. The only difference here is that we have not made the identification $G \times \mathfrak{t}_0^* \cong G \times \mathfrak{t}_0$ that was convenient in our applications to perturbation theory.

10.12 Corollary. *Let $(P, \omega, G, \mathbf{J})$ be a Hamiltonian G-space on which G is acting freely, and for which $\mathbf{J}(P) \subset \mathfrak{g}^*_{\mathrm{reg}}$. Assume the space is geometrically integrable (i.e., $\dim P = \dim G + \operatorname{rank} G$), that P is connected, and that each fiber of $\mathbf{J} : P \to \mathbf{J}(P)$ is compact or has a finite number of connected components. Assume that $U \equiv i(\mathbf{J}(P) \cap \mathcal{W})$ (which is open in $\mathfrak{t}_0^*$) is smoothly contractible. Then the spaces $(P, \omega, G, \mathbf{J})$ and $(G \times U, \omega_G^*, G, \mathbf{J}^G)$ are equivalent, i.e., P admits G-compatible action-group coordinates.*

10.13 Remark. Since $\mathbf{J}(P) \subset \mathfrak{g}^*_{\mathrm{reg}}$ is G-invariant, $i(\mathbf{J}(P) \cap \mathcal{W}) = \mathbf{j}(P)$, where $\mathbf{j} : P \to \mathfrak{t}^*$ is the map defined by 8.7. This fact is useful in computations.

PROOF OF 10.12. Let $(F, \omega_F, T, \mathbf{J}^F)$ denote the symplectic cross-section. Since G acts freely on P and $(P, \omega, G, \mathbf{J})$ is geometrically integrable, T acts freely on F and $(F, \omega_F, T, \mathbf{J}^F)$ is geometrically integrable, by Lemma 10.6.1. By the definition of $\mathbf{J}^F$, we have $i(\mathbf{J}(P) \cap \mathcal{W}) = \mathbf{J}^F(F)$, so that $U = \mathbf{J}^F(F)$ is contractible by hypothesis. By Lemma 10.6.2 and the hypothesis on the fibers of $\mathbf{J}$, each fiber of $\mathbf{J}^F$ is compact or has a finite number of connected components. The hypotheses of Theorem 10.9 therefore apply to the symplectic , which therefore admits T-compatible action-angle coordinates. Theorem 10.2 finishes the proof off. □

CHAPTER 11

The Axisymmetric Euler-Poinsot Rigid Body

In this chapter we apply the general results of the preceding chapter to a nontrivial example, the Euler-Poinsot rigid body. By definition, this is a rigid body fixed at but free to rotate about a point O that is motionless in some inertial frame of reference. It is well-known that the dynamics of such a body is described by integral curves of an appropriate Hamiltonian vector field on $\mathrm{T}^*\,\mathrm{SO}(3)$. We refer the reader to, e.g., Marsden (1992, p. 87) for details. Note that one ordinarily identifies the phase space $\mathrm{T}^*\,\mathrm{SO}(3)$ with $\mathrm{T}\,\mathrm{SO}(3)$, using an appropriate invariant metric (see below).

Let $\boldsymbol{\lambda}, \boldsymbol{\rho} : \mathrm{SO}(3)\times\mathbb{R}^3 \xrightarrow{\sim} \mathrm{T}\,\mathrm{SO}(3)$ denote the usual left and right trivializations:

$$\boldsymbol{\lambda}(\Lambda, m) \equiv \frac{d}{dt}\Lambda e^{t\hat{m}}\Big|_{t=0} \ ,$$

$$\boldsymbol{\rho}(\Lambda, n) \equiv \frac{d}{dt}e^{t\hat{n}}\Lambda\Big|_{t=0} \ .$$

Here $\xi \mapsto \hat{\xi} : \mathbb{R}^3 \to \mathfrak{so}(3)$ is the isomorphism defined by $\hat{\xi}u = \xi \times u$ $(u \in \mathbb{R}^3)$. We think of tangent vectors as equivalence classes of curves and use '$\frac{d}{dt}\Lambda_t|_{t=0}$' to denote the class represented by $t \mapsto \Lambda_t$. In what follows, the vector m will correspond to the *body* angular momentum of the body about O, while n will correspond to the *spatial* angular momentum about O. Note that

11.1 $$\boldsymbol{\lambda}(\Lambda, m) = \boldsymbol{\rho}(\Lambda, n) \Leftrightarrow n = \Lambda m \ .$$

Define a one-form Θ on $\mathrm{T}\,\mathrm{SO}(3)$ as follows. An arbitrary vector tangent to $\mathrm{T}\,\mathrm{SO}(3)$ is of the form $\frac{d}{dt}\boldsymbol{\rho}(\Lambda_t, n_t)|_{t=0}$. Let n' be the element of $\mathbb{R}^3$ uniquely determined by

$$\frac{d}{dt}\Lambda_t\Big|_{t=0} = \boldsymbol{\rho}(\Lambda_0, n') \ .$$

Then by decree,

11.2 $$\langle \Theta, \frac{d}{dt}\boldsymbol{\rho}(\Lambda_t, n_t)\Big|_{t=0}\rangle \equiv n_0 \cdot n' \ ,$$

where the dot denotes the usual Euclidean dot product.

An invariant Riemannian metric on SO(3) is given by

$$\langle\langle \boldsymbol{\rho}(\Lambda, n_1), \boldsymbol{\rho}(\Lambda, n_2)\rangle\rangle \equiv n_1 \cdot n_2 \ .$$

The differential form Θ above is just the canonical one-form on $\mathrm{T}^*\,\mathrm{SO}(3)$ (see Chap. 9), viewed as a one-form on $\mathrm{T}\,\mathrm{SO}(3)$ using the $\langle\langle\,\cdot\,,\,\cdot\,\rangle\rangle$-induced identification $\mathrm{T}^*\,\mathrm{SO}(3) \cong \mathrm{T}\,\mathrm{SO}(3)$. In particular, $\omega = -d\Theta$ is a symplectic two-form on $\mathrm{T}\,\mathrm{SO}(3)$.

Define $H : \mathrm{T}\,\mathrm{SO}(3) \to \mathbb{R}$ by

$$H(\boldsymbol{\lambda}(\Lambda, m)) \equiv \frac{1}{2I_1}m_1^2 + \frac{1}{2I_2}m_2^2 + \frac{1}{2I_3}m_3^2 \ .$$

Then:

The dynamics of an Euler-Poinsot rigid body with moments of inertia I_1, I_2, I_3 about the fixed point O is described by the Hamiltonian system $(\mathrm{T\,SO}(3), \omega, H)$ above.

For a proof, see op. cit.

Assume that the moment of inertia ellipsoid is axisymmetric, i.e., that two of the moments of inertia are equal: $I_1 = I_2 \equiv I$, say. In that case the Hamiltonian is invariant with respect to the following action of $G \equiv \mathrm{SO}(3) \times S^1$ on $\mathrm{T\,SO}(3)$: for $(A, \theta) \in \mathrm{SO}(3) \times S^1$, define

$$(A, \theta) \cdot \boldsymbol{\lambda}(\Lambda, m) \equiv \boldsymbol{\lambda}(A\Lambda R^3_\theta, R^3_\theta m) \ ,$$

where $R^3_\theta \equiv \exp(\theta \hat{e}_3)$ is the rotation about the e_3-axis through angle θ. Alternatively, we may write this action as

$$(A, \theta) \cdot \boldsymbol{\rho}(\Lambda, n) = \boldsymbol{\rho}(A\Lambda R^3_\theta, An) \ .$$

The SO(3) part of the symmetry group corresponds to the familiar rotational 'spatial' symmetry of the system, while the S^1 action corresponds to the axisymmetry of the inertia ellipsoid (i.e., a 'body' symmetry). The action is Hamiltonian with equivariant momentum map $\mathbf{J} : \mathrm{T\,SO}(3) \to \mathfrak{g}^* \cong \mathbb{R}^3 \times \mathbb{R}$ given by

$$\mathbf{J}(\boldsymbol{\lambda}(\Lambda, m)) \equiv (\Lambda m, m_3) \ ,$$

where $m_3 \equiv m \cdot e_3$. Alternatively,

$$\mathbf{J}(\boldsymbol{\rho}(\Lambda, n)) = (n, (\Lambda^{-1} n) \cdot e_3) \ .$$

These formulas are consistent with Noether's theorem (stating that $\mathbf{J}$ is constant on solution curves) and the familiar fact that in an axisymmetric rigid body the spatial angular momentum and the component of body angular momentum along the symmetry axis are conserved.

Our objective is to construct G-compatible action-group coordinates in the space
$(\mathrm{T\,SO}(3), \omega, G, \mathbf{J})$.

Existence

Before constructing action-group coordinates explicitly, we convince ourselves that such coordinates must exist. In the first place, we claim:

$\mathbf{J}$ *has* **compact** *fibers.*

Indeed, suppose that $y \equiv (\mu, a) \in \mathbf{J}(P) \subset \mathbb{R}^3 \times \mathbb{R}$. Then certainly

$$\mathbf{J}^{-1}(y) \subset \boldsymbol{\lambda}\left(\mathrm{SO}(3) \times \overline{B_{\|\mu\|}(0)} \right) \ ,$$

where $\overline{B_r(0)} \subset \mathbb{R}^3$ denotes the closed ball of radius r. Since $\boldsymbol{\lambda}$ is a diffeomorphism, this shows that $\mathbf{J}^{-1}(y)$ is contained in a compact set. Since $\mathbf{J}^{-1}(y)$ is closed in $\mathrm{T\,SO}(3)$ (by the continuity of $\mathbf{J}$), $\mathbf{J}^{-1}(y)$ itself must be compact.

So that G acts freely and the image of the momentum map is contained in $\mathfrak{g}^*_{\mathrm{reg}} = \mathbb{R}^3 \backslash \{0\} \times \mathbb{R}$, we restrict attention to the open dense G-invariant set $P \subset \mathrm{T\,SO}(3)$ defined by

$$P \equiv \{\boldsymbol{\lambda}(\Lambda, m) \mid \Lambda \in \mathrm{SO}(3) \text{ and } m \in \mathbb{R}^3 \text{ and } (m_1 \neq 0 \text{ or } m_2 \neq 0)\} \ .$$

That is, we remove from phase space those points whose body angular momentum vector m lies on the e_3-axis.

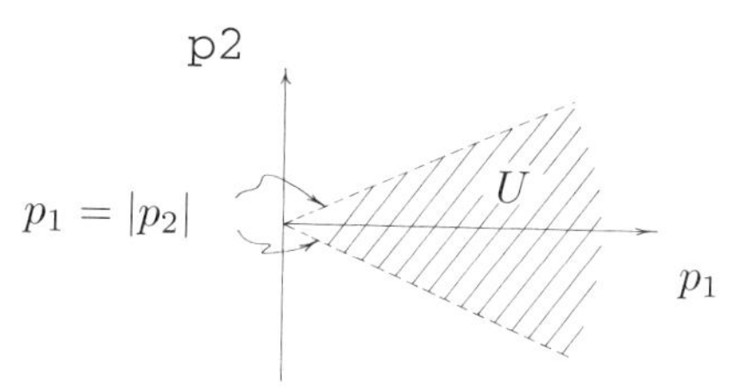

FIGURE 1. The set $U \equiv i(\mathbf{J}(P) \cap \mathcal{W}) = \mathbf{j}(P) = \mathbf{J}^F(F)$ in the axisymmetric Euler-Poinsot rigid body.

For a maximal torus $T \subset G$ we choose

$$T \equiv \{(R^3_\phi, \theta) \in \mathrm{SO}(3) \times S^1 \mid \phi \in [0, 2\pi), \theta \in S^1\} \cong \mathbb{T}^2 \ .$$

Then $\underline{\mathfrak{t}} = \mathrm{span}\{e_3\} \times \mathbb{R}$. For a Weyl chamber $\mathcal{W}$ in $\mathfrak{g}^*$ choose

$$\mathcal{W} \equiv \{(te_3, \tau) \in \mathbb{R}^3 \times \mathbb{R} \mid t \in (0, \infty), \tau \in \mathbb{R}\} \ .$$

For a regular co-adjoint orbit choose $\mathcal{O} \equiv S^2 \times \{0\}$. The unique intersection point of $\mathcal{W}$ and $\mathcal{O}$ is $\mu_0 \equiv (e_3, 0)$.

The projection $\pi_{\mathcal{O}} : \mathbb{R}^3 \backslash \{0\} \times \mathbb{R} \to \mathcal{O} \cong S^2$ is given by $\pi_{\mathcal{O}}(\mu, a) = \mu / \|\mu\|$. The projection $\pi_{\mathcal{W}} : \mathbb{R}^3 \backslash \{0\} \times \mathbb{R} \to \mathcal{W} \cong (0, \infty) \times \mathbb{R}$ is given by $\pi_{\mathcal{W}}(\mu, a) \equiv (\|\mu\|, a)$.

Now $\dim P = 6 = 4 + 2 = \dim G + \operatorname{rank} G$. So $(P, \omega, G, \mathbf{J})$ is geometrically integrable. Since $\mathbf{J}$ has compact fibers, we can apply Corollary 10.12, provided we can show that $U \equiv i(\mathbf{J}(P) \cap \mathcal{W})$ is smoothly contractible. By Remark 10.13, $U = \mathbf{j}(P)$, where $\mathbf{j} \equiv i \circ \pi_{\mathcal{W}} \circ \mathbf{J}$. With our identifications, i is just the identity, so that we obtain $\mathbf{j}(\boldsymbol{\lambda}(\Lambda, m)) = (\|m\|, m_3)$. It follows that

$$U = \{(p_1, p_2) \in (0, \infty) \times \mathbb{R} \mid |p_2| < p_1\}$$

(see Fig. 1). In particular, U is indeed contractible. By Corollary 10.12, G-compatible action-group coordinates exist.

The symplectic cross-section

Since P admits G-compatible action-group coordinates, the symplectic cross-section F admits T-compatible action-angle coordinates (Theorem 10.2). We shall construct action-group coordinates explicitly by applying Proposition 10.7. To this end, we now construct a concrete realization of F, its symplectic structure ω_F, and the momentum map $\mathbf{J}^F : F \to U$.

The symplectic fibration $\pi : P \to \mathcal{O} \cong S^2$ (see Theorem 8.8) is given by $\pi(\boldsymbol{\rho}(\Lambda, n)) = n / \|n\|$. So the symplectic cross-section is

$$\begin{aligned} F \equiv \pi^{-1}(e_3) &= \{\boldsymbol{\rho}(\Lambda, n) \mid \Lambda \in \mathrm{SO}(3) \text{ and } m \in \mathbb{R}^3 \\ &\qquad \text{and } (\Lambda^{-1} n \cdot e_1 \neq 0 \text{ or } \Lambda^{-1} n \cdot e_2 \neq 0) \text{ and } n / \|n\| = e_3\} \\ &= \{\boldsymbol{\rho}(\Lambda, p_1 e_3) \mid \Lambda \in \mathrm{SO}(3) \backslash Z, \ p_1 > 0\} \ , \\ \text{where} \quad & Z \equiv \{\Lambda \in \mathrm{SO}(3) \mid |\Lambda e_3 \cdot e_3| = 1\} \ . \end{aligned}$$

Notice that if one constructs Euler angles on $\mathrm{SO}(3)$, according to the appropriate sign convention, then $Z \subset \mathrm{SO}(3)$, which is just the disjoint union of two circles, corresponds to the coordinate singularities. In particular, if we write $\mathrm{SO}(3)' \equiv \mathrm{SO}(3) \backslash Z$, then $\mathrm{SO}(3)' \cong \mathbb{T}^2 \times (0, \pi)$.

The map $\psi : \mathrm{SO}(3)' \times (0,\infty) \to F$ defined by $\psi(\Lambda, p_1) \equiv \boldsymbol{\rho}(\Lambda^{-1}, p_1 e_3)$ is a diffeomorphism onto F. Let $i_F : F \hookrightarrow P$ denote the inclusion. Then $\psi^* \omega_F = -d\Theta'$, where $\Theta' \equiv \psi^* i_F^* \Theta$. A vector tangent to $\mathrm{SO}(3)' \times (0,\infty)$ is of the form $\frac{d}{dt}(\Lambda \exp(t\hat{\xi}), p_1 + t\tau)\big|_{t=0}$. One computes using our earlier definition of Θ (equation 11.2),

$$11.3 \quad \langle \Theta', \frac{d}{dt}(\Lambda \exp(t\hat{\xi}), p_1 + t\tau)\big|_{t=0}\rangle = \langle \Theta, \frac{d}{dt}\boldsymbol{\rho}(\exp(-t\hat{\xi})\Lambda^{-1}, (p_1 + t\tau)e_3)\big|_{t=0}\rangle = -p(e_3 \cdot \xi) \ .$$

The map $\mathbf{J}^F : F \to \mathbb{R}^2$ is given by $\mathbf{J}^F(\boldsymbol{\rho}(\Lambda, n)) = (n_3, (\Lambda^{-1}n)\cdot e_3)$ for all $\boldsymbol{\rho}(\Lambda, n) \in F$. So $(\mathbf{J}^F \circ \psi)(\Lambda, p_1) = (p_1, (\Lambda e_3 \cdot e_3)p_1)$. As in the proof of 10.12, $\mathbf{J}^F(F) = U$ (computed above).

Let $T \cong \mathbb{T}^2$ act on $\mathrm{SO}(3) \times (0,\infty)$ in the way that makes ψ a $\mathbb{T}^2$-equivariant map. (We will not need to compute this action explicitly.) Then ψ establishes an equivalence between the space

$$(\mathrm{SO}(3)' \times (0,\infty), -d\Theta', \mathbb{T}^2, \mathbf{J}') \ ,$$

and the symplectic cross-section $(F, \omega_F, \mathbb{T}^2, \mathbf{J}^F)$, where Θ' is the one-form computed in 11.3, and $\mathbf{J}' \equiv \mathbf{J}^F \circ \psi : \mathrm{SO}(3)' \times (0,\infty) \to U$ is given by $\mathbf{J}'(\Lambda, p_1) = (p_1, (\Lambda e_3 \cdot e_3)p_1)$.

Explicit action-group coordinates

To construct action-group coordinates by applying 10.7, it remains to exhibit a Lagrangian section for $\mathbf{J}^F$.

Define $s : U \to SO(3)' \times (0,\infty)$ by

$$s(p_1, p_2) \equiv (R^1_{\arccos(p_2/p_1)}, p_1) \ ,$$

where $R^1_\alpha \equiv \exp(\alpha \hat{e}_1)$ is the rotation about the e_1-axis through angle α. By convention, arccos is to take values in $(0, \pi)$.

One verifies immediately that $\mathbf{J}' \circ s = \mathrm{id}_U$, i.e., s is a section of $\mathbf{J}' : \mathrm{SO}(3)' \times (0,\infty) \to U$. We claim that s is in fact Lagrangian. Indeed, for any $\tau_1, \tau_2 \in \mathbb{R}$ we have

$$\langle s^*\Theta', \frac{d}{dt}(p_1 + t\tau_1, p_2 + t\tau_2)\Big|_{t=0}\rangle = \langle \Theta', \frac{d}{dt}(R^1_{\arccos(\frac{p_2 + t\tau_2}{p_1 + t\tau_1})}, p_1 + t\tau_1)\Big|_{t=0}\rangle$$
$$= \langle \Theta', \frac{d}{dt}(R^1_{\arccos(p_2/p_1)} \exp(tf(p_1, p_2, \tau_1, \tau_2)\hat{e_1}), p_1 + t\tau_1)\Big|_{t=0}\rangle \ ,$$

for some real-valued function f. It is irrelevant what f actually is since, applying the formula 11.3 for Θ', we obtain

$$\langle \Theta', \frac{d}{dt}(p_1 + t\tau_1, p_2 + t\tau_2)\Big|_{t=0}\rangle = p_1[e_3 \cdot f(p_1, p_2, \tau_1, \tau_2)e_1] = 0 \ .$$

Since τ_1 and τ_2 were arbitrary, $s^*\Theta' = 0$, implying $s^*(-d\Theta') = 0$. So s is indeed Lagrangian. Since $\psi : \mathrm{SO}(3)' \times (0,\infty) \to F$ is an equivalence, $\psi \circ s : U \to F$ is a Lagrangian section for $\mathbf{J}^F : F \to U$. One computes

$$(\psi \circ s)(p_1, p_2) \equiv \boldsymbol{\rho}\left(R^1_{-\arccos(p_2/p_1)}, p_1 e_3\right) \ .$$

We may now apply Proposition 10.7 as follows: Define $\phi : G\times U = SO(3)\times S^1\times U \to P \subset \mathrm{T}SO(3)$ by

$$\begin{aligned}\phi((A,\theta),(p_1,p_2)) &\equiv (A,\theta)\cdot((\psi\circ s)(p_1,p_2))\\ &= \boldsymbol{\rho}(AR^1_{-\arccos(p_2/p_1)}R^3_\theta,\ p_1Ae_3)\\ &= \boldsymbol{\lambda}(AR^1_{-\arccos(p_2/p_1)}R^3_\theta,\ p_1R^3_{-\theta}R^1_{\arccos(p_2/p_1)}e_3)\\ &= \boldsymbol{\lambda}(AR^1_{-\arccos(p_2/p_1)}R^3_\theta,\ m_{p_1,p_2,\theta})\ ,\end{aligned}$$

$$\text{where}\quad m_{p_1,p_2,\theta} \equiv (-\sqrt{p_1^2-p_2^2}\sin\theta, -\sqrt{p_1^2-p_2^2}\cos\theta, p_1)\ .$$

Then ϕ is an equivalence realizing G-compatible action-group coordinates in the space P. The Hamiltonian ϕ^*H pulled back to $G\times U$ is given by $\phi^*H(g,p) = h(p)$, where

$$h(p_1,p_2) \equiv \frac{1}{2I}p_1^2 + \frac{1}{2}(\frac{1}{I_3} - \frac{1}{I})p_2^2\ .$$

Notice that h is convex (in the sense of Definition 5.3, Part 1) in the *prolate* ($I_3 < I$) case.

CHAPTER 12

Passing from Dynamic Integrability to Geometric Integrability

Let P be a Hamiltonian G-space with G acting freely and $\mathbf{J}(P) \subset \mathfrak{g}^*_{\text{reg}}$; see p. 15. Assume the space is dynamically integrable (see 3.7) and let $H : P \to \mathbb{R}$ be a G-invariant Hamiltonian. In this chapter we address the problem of enlarging the symmetry group G of H to $G \times S^1$, in such a way as to render the space *geometrically* integrable. (Of course we can always extend the symmetry group by a factor $\mathbb{R}$ by considering the flow of X_H, assuming that X_H is complete. The point is that we seek an extension that is *compact*.) Such a system is then amenable to the action-group coordinate constructions described in Chap. 10.

We have already shown (Lemma 10.6) that P is geometrically integrable if and only if its symplectic cross-section F is geometrically integrable. An analogous statement holds for dynamic integrability, which one proves in the same way. With Lemma 8.17 in hand one can easily reduce the above S^1 extension problem to the Abelian case by passing to the symplectic cross-section. We therefore restrict attention to the case of a *toral symmetry*:

12.1 Assumption. In this chapter T denotes a torus, $(F, \omega, T, \mathbf{J})$ a Hamiltonian T-space on which T is acting freely, and $H : F \to \mathbb{R}$ a fixed T-invariant Hamiltonian. We suppose that $(F, \omega, T, \mathbf{J})$ is **dynamically** integrable, i.e., $\dim F = 2 \dim T + 2$.

Our task then is to extend the T-action on F to a $T \times S^1$ action with respect to which H is invariant.

We do not attempt great generality but focus on the case in which the reduced dynamics of H in each (two-dimensional) reduced space is composed of *an equilibrium point surrounded by periodic trajectories*. These reduced spaces may be realized as symplectic leaves of F/T (or almost; see below) and the periodic orbits then coincide (as sets) with the orbits of an S^1-action on F/T. Our strategy is to discover how to lift this action to an S^1-action on F commuting with that of T. Our final conclusions are formulated in Theorem 12.6 on p. 72.

In Appendix D we illustrate the general theory developed here in the Euler-Poinsot rigid body. (This includes a concrete application of Lemma 8.17.)

Preliminaries

Let $\rho : F \to F/T$ denote the natural projection. The Poisson structure of F drops to one on F/T (see, e.g., Marsden (1992)). Since T is Abelian and we assume momentum maps are Ad^*-equivariant, $\mathbf{J} : F \to \mathfrak{t}^*$ is actually T-*in*variant. This implies the existence of a submersion $\mathbf{j} : F/T \to \mathfrak{t}^*$ with $\mathbf{J} = \mathbf{j} \circ \rho$ whose image is $U \equiv \mathbf{J}(P)$. We call the fibers $F_\mu \equiv \mathbf{j}^{-1}(\mu)$ $(\mu \in U)$ of $\mathbf{j}$ *pseudoleaves*. Like the symplectic leaves of F/T, each fiber F_μ is an integral manifold of the characteristic distribution of F/T supporting a symplectic structure ω_μ, with respect to which the inclusion $F_\mu \hookrightarrow F/T$ is Poisson. The connected components of F_μ are bona fide symplectic leaves; if $\mathbf{J} : F \to U$ has connected fibers then the pseudoleaves and symplectic leaves coincide. Each leaf[1] F_μ is symplectomorphic to the Marsden-Weinstein reduced space $\mathbf{J}^{-1}(\mu)/F$ (see 3.6, p. 18). These facts are corollaries of Theorem E.18, Appendix E.

Obstructions to lifting S^1 symmetries

12.2 Definition. A function $I : P \to \mathbb{R}$ on a Poisson manifold P is called a *circle action generator* if it is the momentum map for a free Hamiltonian action of S^1 on P, i.e., if all orbits of X_I are minimally 2π-periodic.

Let I be a circle action generator on an open set $V \subset F/T$. We seek to lift the corresponding S^1-action to one on $\rho^{-1}(V) \subset F$ by taking as generator the candidate $I \circ \rho$. In general $I \circ \rho$ is *not* a generator however, an obstruction being the existence of non-trivial *reconstruction phases*: If X_h is a Hamiltonian vector field on F/T with a t_1-periodic orbit $t \mapsto y(t)$, and if $t \mapsto x(t) \in F$ is one of the corresponding orbits of $X_{h\circ\rho}$ covering $t \mapsto y(t)$, then the *reconstruction phase* of the orbit $t \mapsto y(t)$ is the unique $q \in T$ such that $x(t_1) = q \cdot x(0)$. The S^1 action generated by I lifts as proposed provided all orbits of X_I have trivial reconstruction phases. Fortunately, we need not check the phase of *every* orbit:

12.3 Theorem. *If $I : V \to \mathbb{R}$ is a circle action generator and $V \cap F_{\mu_0}$ is connected, then all X_I-orbits lying in the leaf F_{μ_0} have identical reconstruction phase.*

To prove 12.3 we first formulate a general formula for reconstruction phases due to Blaom (2000). (In fact we use a slight variation of Blaom's formula, established by adapting his proof in a more-or-less obvious way.) Fix some smooth distribution D on F/T furnishing a complement in $\mathrm{T}(F/T)$ for the characteristic distribution (the distribution tangent to the symplectic leaves). Then for each $y \in F/T$ there is an isomorphism $L_y : \mathfrak{t}^* \to D(y)$ whose inverse is $v \mapsto \langle d\mathbf{j}, v\rangle$. If λ is an $\mathbb{R}$-valued k-form on F/T, then there is a $\mathfrak{t}$-valued k-form $D_\mu\lambda$ on F_μ defined through

$$\langle \nu, D_\mu\lambda(v_1, \dots, v_k)\rangle = d\lambda(L_y(\nu), v_1, \dots, v_k) \qquad (\nu \in \mathfrak{t}^*,\ v_j \in \mathrm{T}_yF_\mu,\ y \in F_\mu) \ .$$

12.4 Theorem. *Let $\Sigma \subset F_\mu$ be a compact oriented 2-manifold with boundary $\partial\Sigma = y_2 - y_1$, where $y_1(\,\cdot\,)$ and $y_2(\,\cdot\,)$ are periodic orbits of some Hamiltonian vector field X_h on F/T with periods Γ_1 and Γ_2. Then the difference in the corresponding reconstruction phases for the two orbits is given by*

$$q_2q_1^{-1} = \exp\left(\int_0^{\Gamma_2} D_\mu h(y_2(t))\,dt - \int_0^{\Gamma_1} D_\mu h(y_1(t))\,dt + \int_\Sigma D_\mu\omega_D\right) ,$$

where ω_D is the unique two-form on F/T whose kernel is D, and whose restriction to a symplectic leaf gives that leaf's symplectic structure.

Proof of 12.3. By the connectedness hypothesis it suffices to demonstrate that the reconstruction phase of orbits is locally constant on $V \cap F_{\mu_0}$. Since $(F, \omega, T, \mathbf{J})$ is dynamically integrable, the fibers of $\mathbf{j}$ are two-dimensional. Define $\sigma : V \to U \times \mathbb{R}$ by $\sigma(y) \equiv (I(y), \mathbf{j}(y))$. Since I is a nondegenerate function on V (X_I has no equilibrium points) and $\mathbf{j}$ is a submersion, σ is a submersion. The

[1] Henceforth we drop the qualifier 'pseudo.'

orbits of X_I are the intersections of I-level sets with symplectic leaves (fibers of $\mathbf{j}$). So the nonempty fibers of σ are the orbits of the free S^1-action generated by I, implying $\sigma : V \to \sigma(V)$ is a locally trivial S^1-bundle. That is, each point in V has a connected neighborhood V' that is the domain for a diffeomorphism ψ making the following diagram commute:

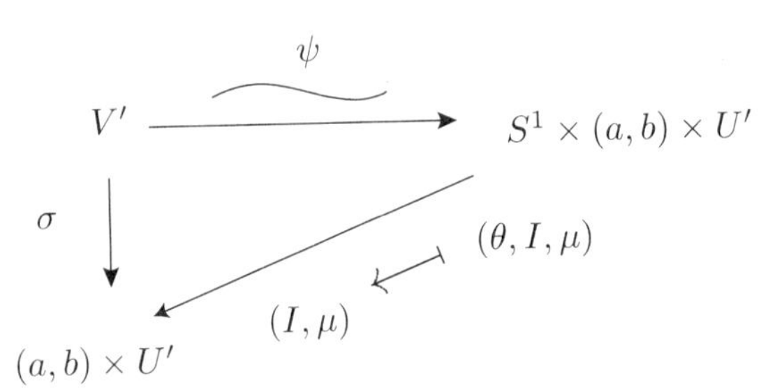

Here $(a,b) \subset \mathbb{R}$ is some open interval and $U' \subset U$ denotes some open set.

We have $\psi(y) = (\theta(y), I(y), j(y))$ $(y \in V')$, for some angular 'function' θ on V'. A leaf $F_\mu = \mathbf{j}^{-1}(\mu)$ intersects the neighborhood V' in a cylinder

12.5 $$F_\mu \cap V' = \psi^{-1}\left(S^1 \times (a,b) \times \{\mu\}\right) \cong S^1 \times (a,b) \ ,$$

coordinatized by the restrictions of θ and I. The action of S^1 generated by I on this cylinder is the standard one on $S^1 \times (a,b)$, under the obvious identification being made in 12.5. *It follows that the restrictions of θ and I to the cylinder amount to local action-angle coordinates for the leaf.* That is, the symplectic structure on the cylinder $F_\mu \cap V'$ is the restriction of $d\theta \wedge dI$.

On V' a distribution D furnishing a complement for the characteristic distribution is defined by

$$D(y) \equiv \left\{\nu \cdot \frac{\partial}{\partial \mu}(y) \mid \nu \in \mathfrak{g}^*\right\} \qquad (y \in V') \ ,$$

where $\nu \cdot \partial/\partial\mu$ $(\nu \in \mathfrak{g}^*)$ is the vector field on V' defined by

$$\nu \cdot \frac{\partial}{\partial \mu}(y) \equiv \frac{d}{dt}\psi^{-1}(\theta, I, \mu + t\nu)\Big|_{t=0} \qquad ((\theta, I, \mu) \equiv \psi(y)) \ .$$

The corresponding two-form ω_D (see 12.4) is given by $\omega_D = d\theta \wedge dI$, implying that $d\omega_D = 0$. In particular, $D_{\mu_0}\omega_D = 0$. One determines that

$$L_y(\nu) = \nu \cdot \frac{\partial}{\partial \mu}(y) \qquad (y \in F_{\mu_0} \cap V',\ \nu \in \mathfrak{t}^*)$$

and consequently that $D_{\mu_0} I = 0$. Applying Theorem 12.4 with $h \equiv I$, $\Sigma \equiv \psi^{-1}(S^1 \times (a,b) \times \{\mu_0\})$ and F replaced by $\rho^{-1}(V')$ finishes the proof. □

The main result

We are now ready to prove the following:

12.6 Theorem. *Let $h : F/T \to \mathbb{R}$ denote the function to which H drops on account of T-invariance. In addition to 12.1, assume the following hold (replacing F with an appropriate T-invariant subset if necessary):*

1. *Each leaf F_μ $(\mu \in U)$ is diffeomorphic to an open disk in $\mathbb{R}^2$.*

2. *There exists a submanifold $Z \subset F/T$ of codimension two that is transversal to each leaf, and having the following property (see Fig. 1): Z intersects F_μ at a single point z_μ that is an equilibrium point of X_h.*
3. *Each punctured disk $F_\mu^\times \equiv F_\mu \backslash \{y_\mu\}$ is foliated by nontrivial periodic orbits of X_h.*

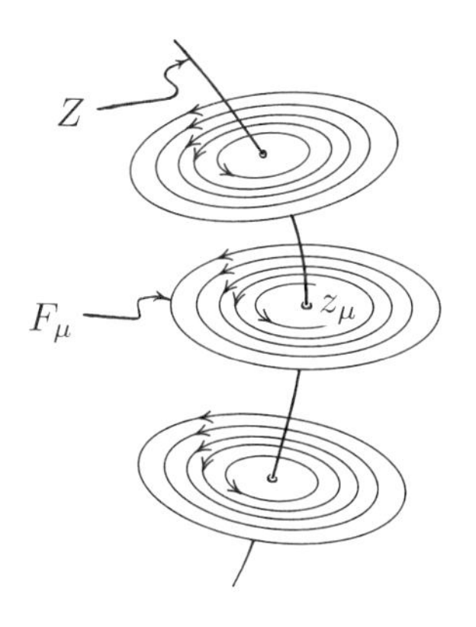

FIGURE 1. The reduced dynamics in F/T, as described in the hypotheses of Theorem 12.6.

Next, let $\mathfrak{t}^{\mathbb{Z}} \subset \mathfrak{t}$ denote the kernel of $\exp : \mathfrak{t} \to T$ *and write* $\frac{1}{2\pi}\mathfrak{t}^{\mathbb{Z}} \equiv \{\frac{1}{2\pi}\nu \mid \nu \in \mathfrak{t}^{\mathbb{Z}}\}$; *let $\xi_0 \in \frac{1}{2\pi}\mathfrak{t}^{\mathbb{Z}}$ and $c_0 \in \mathbb{R}$ be arbitrary, and define $I : F/T \to \mathbb{R}$ by*

$$I(y) \equiv c_0 + \langle \mathbf{j}(y), \xi_0 \rangle + \frac{1}{2\pi} \int_{\Pi_y} \omega_{\mathbf{j}(y)} \ ,$$

where $\Pi_y \subset F_{\mathbf{j}(y)}$ denotes the oriented two-manifold whose boundary is the X_h-orbit through y. Then:

4. *The orbits of $X_{I\circ\rho}$ in $F^\times \equiv F\backslash Z$ are minimally 2π-periodic.*
5. *The action of S^1 on F generated by $X_{I\circ\rho}$ commutes with that of T.*
6. *The Hamiltonian H is S^1-invariant.*
7. *The map $I \circ \rho : F^\times \to \mathbb{R}$ is both S^1- and T-invariant.*
8. *The momentum map $\mathbf{J} : F^\times \to \mathfrak{t}^*$ is S^1-invariant.*

In particular, the action of T extends to a Hamiltonian action by $T \times S^1$ having $T \times S^1$-invariant momentum map $\mathbf{J} \times I \circ \rho$, where $(\mathbf{J} \times I \circ \rho)(x) \equiv (\mathbf{J}(x), (I \circ \rho)(x))$. Furthermore, H is $T \times S^1$-invariant, $T \times S^1$ acts freely on $F^\times$, and the space $(F^\times, \omega, T \times S^1, \mathbf{J} \times I \circ \rho)$ is ***geometrically*** *integrable.*

12.7 Remark. The choice of constant c_0 is immaterial. However, different choices of $\xi_0 \in \frac{1}{2\pi}\mathfrak{t}^{\mathbb{Z}}$ lead to genuinely different S^1-actions on $F^\times$, although the corresponding $T \times S^1$-orbits are always the same. We examine this non-uniqueness in the axisymmetric rigid body in Appendix D.

PROOF OF 12.6. The restriction of I to a leaf F_μ is given by

$$I|F_\mu = \text{constant} + \frac{1}{2\pi} \int_{\Pi_y} \omega_\mu \ .$$

Reminded of classical action-angle coordinate constructions in the plane (see, e.g., Melnikov (1963)), we conclude that I is a circle action generator on $F_\mu^\times$ in the sense of 12.2. Since $\mu \in U$ is arbitrary, I is a circle action generator on $(F/T)^\times \equiv (F/T)\setminus Z$. By Condition 3 and the definition of I it is not difficult to see that the level sets of I and h, in each leaf F_μ, coincide. Consequently h is invariant with respect to the S^1-action on $(F/T)^\times$ generated by I.

To prove Part 4 of the theorem we must show that all the orbits of X_I in $(F/T)^\times$ have trivial reconstruction phase. Fixing $\mu \in U$, we already know from Theorem 12.3 (taking $V \equiv (F/T)^\times$) that all X_I-orbits in $F_\mu^\times$ have a common reconstruction phase $q_\mu \in T$, say. To prove $q_\mu = \mathrm{id}$, we first study the dynamics of $X_{I\circ\rho}$ on the 'singular' invariant torus $\rho^{-1}(z_\mu)$, and then apply an appropriate limiting argument. To this end, we require the following elementary corollary of Blaom (2000, Theorem B):

12.8 Proposition. *Let $I : F/T \to \mathbb{R}$ be an arbitrary function and assume X_I has an equilibrium point $z \in F_\mu$. Then at all x on the T-orbit $\rho^{-1}(z)$ one has*

$$X_{I\circ\rho}(x) = \xi_F(x) \ , \quad \text{where } \xi \equiv D_\mu I(z)$$

and ξ_F denotes infinitesimal generator.

Here D_μ is the operator defined on p. 71. However, as $d_z I \in \operatorname{Ann} \mathrm{T}_z F_\mu$ (because $X_I(z) = 0$), we may describe $D_\mu I(z)$ in a simpler way: Let $\tilde{I}$ be a function on $\mathfrak{t}^*$ defined in a neighborhood of μ such that $d_z I = d_z(\tilde{I} \circ \mathbf{j})$; then $D_\mu I(z) = \frac{\delta \tilde{I}}{\delta \mu}$. If I is the function defined in 12.6 and we take $z \equiv z_\mu$, then we may choose $\tilde{I}(\mu') \equiv c_0 + \langle \xi_0, \mu' \rangle$ $(\mu' \in \mathfrak{t}^*)$, implying $D_\mu I(z_\mu) = \xi_0$. It follows from 12.8 that the flow map $(t, x) \mapsto \Phi_t(x)$ of $X_{I\circ\rho}$ satisfies

$$\Phi_t(x) = \exp(t\xi_0) \cdot x \quad \text{whenever} \quad \rho(x) = z_\mu \ . \tag{12.9}$$

Let $\tau \mapsto x_\tau : [0, 1] \to F$ be any continuous curve such that $\rho(x_\tau) \in F_\mu^\times$ for $\tau \neq 1$ and $\rho(x_1) = z_\mu$. Such a curve certainly exists since $\rho : F \to F/T$ is a locally trivial fiber bundle (T acts freely). We have

$$\begin{aligned} \Phi_{2\pi}(x_\tau) &= q_\mu \cdot x_\tau \qquad (\tau \neq 1) \ , \\ \Phi_{2\pi}(x_1) &= x_1 \ . \end{aligned}$$

The first equation follows from the definition of reconstruction phases, the second from 12.9 and the assumption $\xi_0 \in \frac{1}{2\pi}\mathfrak{t}^{\mathbb{Z}}$. Taking the limit $\tau \to 1$ in the first equation and applying the continuity of solutions with respect to initial data gives $\Phi_{2\pi}(x_1) = q_\mu \cdot x_1$. By the second equation we therefore have $q_\mu \cdot x_1 = x_1$. Since T acts freely, $q_\mu = \mathrm{id}$.

Since $X_{I\circ\rho}$ is evidently T-invariant, its flow is T-equivariant. This proves Part 5. Part 6 follows from the forementioned S^1-invariance of h, and the S^1-equivariance of $\rho : F^\times \to (F/T)^\times$, which in turn follows from the ρ-relatedness of $X_{I\circ\rho}$ and X_I. That $I \circ \rho$ is T-invariant is trivial. That it is S^1-invariant follows from the same argument applied earlier to $H = h \circ \rho$. This establishes Part 7. Part 8 follows from the S^1-equivariance of ρ, the X_I-invariance of the leaves F_μ $(\mu \in U)$, and the formula $\mathbf{J} = \mathbf{j} \circ \rho$. That $T \times S^1$ acts freely on $F^\times$ follows from the following facts: (i) T acts freely on F; (ii) $\rho : F^\times \to (F/T)^\times$ is S^1-equivariant (see above); and (iii) S^1 acts freely on $(F/T)^\times$. The other claims of the theorem are straightforward. □

CHAPTER 13

Concluding Remarks

Extensions to non-free actions and non-regular momentum values

We conjecture that the estimates on the evolution of $\mathbf{J}(x_t)$ described in the Introduction (and in detail in Chap. 7) hold under weaker hypotheses than the existence of action-group coordinates requires. Of course one still needs some kind of geometric integrability for the unperturbed system (see p. 18), and one may need to assume uniform orbit and co-adjoint orbit types (see p. 16). Also, a convexity (or quasi-convexity) assumption will be necessary. To generalize our results to other orbit and co-adjoint orbit types will require more general Hamiltonian G-space normal forms. Moreover, one will need very concrete realizations of these spaces, as well as a good handle on the symplectic structure, equations of motion, etc. In addition, one will need explicit (local) complexifications of the space and its associated structures. To be useful in applications one will require *constructive* ways of realizing these spaces in examples.

One rather general Hamiltonian G-space normal form is the 'Marle-Guillemin-Sternberg' normal form described in, e.g., Ortega (1998), suitably specialized to the integrable case. Already some of the criteria mentioned above are met, a notable exception being availability of constructive existence proofs. In any case, as the normal form involves certain associated bundles ('twisted products'), it is probably too complicated to be considered yet in full generality. However, under suitable restrictions (but weaker than those necessary for action-group coordinates) one would expect these bundles to trivialize, giving a space to which Theorem 5.8 (p. 30) already applies. Rather than attempt to formulate such conditions here, we now relate two examples where this appears to be the case.

Recall that Theorem 5.8 applies to 'coordinates' of the form $G \times U$, for some open subset U of a vector space $\mathfrak{t}$ and some compact real-analytic manifold G. For example, G could be a quotient of groups G_1/G_2. A complexification of G in that case would be $G_1^{\mathbb{C}}/G_2^{\mathbb{C}}$. An example is geodesic motions on S^3 (relevant to the regularized Kepler problem). The phase space is T^*S^3 and a symmetry group is $\mathrm{SO}(4)$, acting by the cotangent lift of the standard action on $S^3 \subset \mathbb{R}^4$. If we remove the zero-section $\mathbf{0}$ from T^*S^3, then all points in $P \equiv \mathrm{T}^*S^3\backslash\mathbf{0}$ have orbit type $(\mathrm{SO}(2))$ and co-adjoint orbit type (T), where $T \subset \mathrm{SO}(4)$ is a maximal torus (so that we still have $\mathbf{J}(P) \subset \mathfrak{g}^*_{\mathrm{reg}}$ but G does not act freely on P). The orbit space $P/\mathrm{SO}(4)$ is identifiable with $(0,\infty)$ and the orbit space projection $P \to (0,\infty)$ is in fact a trivial bundle: $P \cong \mathrm{SO}(4)/\mathrm{SO}(2) \times (0,\infty)$. So we may take $G_1 \equiv \mathrm{SO}(4)$, $G_2 \equiv \mathrm{SO}(2)$, and $U \equiv (0,\infty)$.

In the 1:1:1 resonance the phase space is $\mathbb{R}^6 \cong \mathbb{C}^3$, and a symmetry group is $\mathrm{SU}(3)$. Removing the origin, one obtains a uniform orbit type $(\mathrm{SU}(2))$ and co-adjoint orbit type $(\mathrm{SU}(2))$, so that *both* our orbit type conditions fail. Nevertheless one has diffeomorphisms $\mathbb{R}^6\backslash\{0\} \cong S^5 \times (0,\infty) \cong \mathrm{SU}(3) / \mathrm{SU}(2) \times (0,\infty)$, which is again of the form proposed above.

In both the above examples one can prove that Assumption A (the existence of a 'period lattice' on p. 27) holds. It may be worth trying to verify the remaining Assumptions B and C. Note that in both examples the space U can be identified with some open *wall* of the Weyl chamber of the symmetry group.

Homogeneous spaces

One would like to apply our Nekhoroshev estimates to geodesic motions on a general homogeneous space G/H (G compact, $H \subset G$ closed), equipped with an almost G-invariant metric. The difficulties in doing so are as follows. Firstly, the orbit type conditions necessary for the existence of action-group coordinates may fail to exist. An example is S^3 ($G = \mathrm{SO}(4)$, $H = \mathrm{SO}(3)$) discussed above. However, as discussed in this example, generalizations of our methods may nevertheless deliver Nekhoroshev estimates. A more serious problem is that G-invariance of the metric may be insufficient to guarantee *geometric* integrability. An example is the Euler-Poinsot rigid body discussed in Chap. 10 ($G = \mathrm{SO}(3)$, $H = \{\mathrm{id}\}$), which is geometrically integrable only once the G-symmetry coming from homogeneity is enlarged to $G \times S^1$-symmetry, by assuming body axisymmetry or, more generally, by carrying out the construction described in Chap. 12. For some other dynamically integrable examples of homogeneous spaces the latter strategy may also succeed, but in general one may have to work hard to find a compact symmetry group large enough to render the unperturbed system geometrically integrable. (As far as we are aware, this is still an open problem in the N-dimensional rigid body, which has long been known to be integrable in the classical sense (Manakov, 1976).) In any case, this 'supersymmetry' will generally satisfy orbit type conditions necessitating the use of normal forms more sophisticated than action-group coordinates (see above).

Non-compact symmetry groups

We do not know the extent to which our results can be generalized to the case of non-compact non-Abelian G, except to say that more restrictive assumptions on the perturbations considered will be necessary, as in the non-compact Abelian case ($G = \mathbb{T}^{n-k} \times \mathbb{R}^k$, $k \neq 0$). Perhaps connected and simply connected nilpotent groups, being diffeomorphic to $\mathbb{R}^n$, would be a simple place to begin.

APPENDIX A

Proof of the Nekhoroshev-Lochak Theorem

This appendix is devoted to proving Theorem 5.8, p. 30. To a large extent we follow Lochak's original arguments (Lochak (1992), Lochak (1993)). With the exception of some comments of our own on the role of convexity, we offer minimal motivation. For deeper insight, the reader is referred to Lochak's original papers, especially Lochak (1993).

To keep our exposition as simple as possible, we have restrained from employing two 'technical tricks' that would have improved our estimates. The first, discovered by Neishtadt (1984) and implemented in Lochak and Neishtadt (1992), concerns the normal form calculation. The idea is that one should make the first coordinate change larger than the others. The second trick we could have used is to rescale the frequency (in the simultaneous approximation argument) so that its largest component is unity. This allows one to lower the dimension by one. See Lochak (1992) for details.

Consider the setup described in Chap. 5 and restrict attention to a value of the parameter p_* with the property that $\Omega(p_*) = \frac{1}{T_*}\mathbf{n}$, for some $T_* > 0$ and $\mathbf{n} \in \mathfrak{t}^{\mathbb{Z}}$ ($\Omega \equiv \nabla h$). Then according to Assumption A, $G^{\bar{\sigma}} \times \{p_*\}$ is foliated by T_*-periodic solution trajectories of X_{H_0} ($H_0(g,p) \equiv h(p)$). One of our first objectives is to derive a normal form for the perturbed Hamiltonian $H = H_0 + F$ in a neighborhood of $G \times \{p_*\}$ (Proposition A.16 below). We begin with some preliminaries:

Averaging over periodic orbits

Define

A.1
$$W(g,p) \equiv w(p) \equiv \Omega(p_*) \cdot p \ ,$$

so that $\nabla w(p) = \frac{1}{T_*}\mathbf{n}$ *for all* $p \in B^{\bar{\rho}}$. By Assumption A, X_W has a well-defined flow map $(t,x) \mapsto \Phi_W^t(x) : \mathbb{R} \times (G^{\bar{\sigma}} \times B^{\bar{\rho}}) \to G^{\bar{\sigma}} \times B^{\bar{\rho}}$ satisfying $\Phi_W^{t+T_*} = \Phi_W^t$, and such that Φ_W^t maps a set of the form $G^{\sigma} \times \{p\}$ ($0 \leqslant \sigma \leqslant \bar{\sigma}$, $p \in B^{\bar{\rho}}$) onto itself. In particular Φ_W^t maps $D_\gamma(p_*,r)$ onto $D_\gamma(p_*,r)$ for any values of the parameters γ and r, and is the identity map when $t = T_*$.

For $u \in \mathcal{A}^{\mathbb{C}}(D_\gamma(p_*,r))$, we define $\bar{u}, \tilde{u}, \mathcal{I}u \in \mathcal{A}^{\mathbb{C}}(D_\gamma(p_*,r))$ by

$$\bar{u} \equiv \frac{1}{T_*}\int_0^{T_*} u \circ \Phi_W^t dt \ ,$$
$$\tilde{u} \equiv u - \bar{u} \ ,$$
$$\mathcal{I}u \equiv \frac{1}{T_*}\int_0^{T_*} tu \circ \Phi_W^t dt \ .$$

Clearly $\|\bar{u}\|_\gamma^{p_*,r} \leqslant \|u\|_\gamma^{p_*,r}$ and $\|\tilde{u}\|_\gamma^{p_*,r} \leqslant 2\,\|u\|_\gamma^{p_*,r}$.

A.2 Lemma.

1. $$u + \{W, \mathcal{I}u\} = \bar{u}$$

2. $$\|\mathcal{I}u\|_\gamma^{p_*,r} \leqslant \frac{1}{2} T_* \|u\|_\gamma^{p_*,r}$$

PROOF. The estimate A.2.2 is obvious. Regarding A.2.1, one computes

$$\begin{aligned}
\{W, \mathcal{I}u\} &= -\frac{d}{d\tau}\mathcal{I}u \circ \Phi_W^\tau\big|_{\tau=0} \\
&= -\frac{1}{T_*}\frac{d}{d\tau}\int_0^{T_*} t(u \circ \Phi_W^t) \circ \Phi_W^\tau dt\big|_{\tau=0} \\
&= -\frac{1}{T_*}\int_0^{T_*} t(\frac{d}{dt}u \circ \Phi_W^t) d\tau \\
&= -\frac{1}{T_*}\left[tu \circ \Phi_W^t\right]_0^{T_*} + \frac{1}{T_*}\int_0^{T_*} u \circ \Phi_W^t dt \\
&= -u + \bar{u}
\end{aligned}$$

□

Lie transforms and the Iterative Lemma

In what follows we shall be interested in Hamiltonians of the form

$$\mathcal{H} \equiv H_0 + Z + N \qquad (Z, N \in \mathcal{A}^{\mathbb{C}}(D_\gamma(p_*, r))) \ ,$$

with $\tilde{Z} = 0$. The Hamiltonian 5.4 is of this form if we define $\mathcal{H} \equiv H$, $Z \equiv \bar{F}$ and $N \equiv \tilde{F}$.

In the method known as *Lie transforms*, one makes symplectic changes of coordinates using the time-one map associated with the Hamiltonian flow generated by some 'auxiliary' Hamiltonian. Indeed let $\chi \in \mathcal{A}^{\mathbb{C}}(D_\gamma(p_*, r))$ be given, and suppose that X_χ has a well-defined flow map $(t, x) \mapsto \Phi_\chi^t(x) : [-1, 1] \times D_{\gamma-\delta}(p_*, r) \to D_\gamma(p_*, r)$, for some $\delta \in (0, \gamma]$. Then $\phi \equiv \Phi_\chi^1$ is symplectic. The following computations make use of Taylor's formula with integral remainder:

$$\begin{aligned}
\mathcal{H} \circ \phi =& (H_0 + Z + N) + (H_0 \circ \phi - H_0) + \big((Z + N) \circ \phi - (Z + N)\big) \\
=& (H_0 + Z) + N + (W \circ \phi - W) \\
& + \big((H_0 - W) \circ \phi - (H_0 - W)\big) + \big((Z + N) \circ \phi - (Z + N)\big) \\
=& (H_0 + Z) + (N + \{W, \chi\}) \\
& + \int_0^1 (1 - \tau)\{\{W, \chi\}, \chi\} \circ \Phi_\chi^\tau d\tau + \Delta_1(H_0 - W) + \Delta_1(Z + N) \ ,
\end{aligned}$$

where

$$\text{A.3} \qquad \Delta_1 u \equiv u \circ \phi - u = \int_0^1 \{u, \chi\} \circ \Phi_\chi^\tau d\tau \qquad \big(u \in \mathcal{A}^{\mathbb{C}}(D_\gamma(p_*, r))\big) \ .$$

Choosing $\chi \equiv \mathcal{I}N$ and applying A.2.1 (with $u \equiv N$), we obtain

$$\begin{aligned}
\mathcal{H} \circ \phi = (H_0 + Z') + \int_0^1 (1 - \tau)\{\{W, \chi\}, \chi\} \circ \Phi_\chi^\tau d\tau \\
+ \Delta_1(H_0 - W) + \Delta_1(Z + N) \ ,
\end{aligned}$$

where $Z' \equiv Z + \bar{N}$ (so that $\tilde{Z}' = 0$). By A.2.1, and our choice $\chi \equiv \mathcal{I}N$,

$$\{W, \chi\} = \bar{N} - N = -\tilde{N} \ ,$$

so that

$$\int_0^1 (1-\tau)\{\{W,\chi\},\chi\} \circ \Phi_\chi^\tau d\tau = -\int_0^1 (1-\tau)\{\tilde{N},\chi\} \circ \Phi_\chi^\tau d\tau \ .$$

We may therefore write $\mathcal{H} \circ \phi = H_0 + Z' + N'$, if we define

$$N' \equiv \Delta_1(H_0 - W) + \Delta_1(Z + N) - \Delta_2 \tilde{N}$$

where

A.4 $$\Delta_2 u \equiv \int_0^1 (1-\tau)\{u,\chi\} \circ \Phi_\chi^\tau d\tau \qquad \left(u \in \mathcal{A}^{\mathbb{C}}(D_\gamma(p_*, r))\right) \ .$$

Summarizing:

A.5 Lemma. *Let $Z, N \in \mathcal{A}^{\mathbb{C}}(D_\gamma(p_*, r))$ and $0 < \delta \leqslant \gamma$ be given, with $\tilde{Z} = 0$, and assume that the Hamiltonian vector field associated with $\chi \equiv \mathcal{I}N$ has a well-defined flow map*

$$(t, x) \mapsto \Phi_\chi^t : [-1, 1] \times D_{\gamma-\delta}(p_*, r) \to D_\gamma(p_*, r) \ .$$

Then the Hamiltonian

$$\mathcal{H} \equiv H_0 + Z + N$$

is pulled back by the symplectic coordinate transformation $\phi \equiv \Phi_\chi^1$ to

$$\mathcal{H} \circ \phi = H_0 + Z' + N' \ ,$$

where

1. $$Z' \equiv Z + \bar{N} \qquad \textit{(so that } \tilde{Z}' = 0) \ ,$$
2. $$N' \equiv \Delta_1(H_0 - W) + \Delta_1(Z + N) - \Delta_2 \tilde{N}$$

and where $\Delta_1, \Delta_2 : \mathcal{A}^{\mathbb{C}}(D_\gamma(p_*, r)) \to \mathcal{A}^{\mathbb{C}}(D_{\gamma-\delta}(p_*, r))$ are the operations defined by equations A.3, A.4.

A.6 Remark. The term $\Delta_1(H_0 - W)$ corresponds to what Lochak (1992) refers to as the 'frequency shift' contribution to the transformed Hamiltonian.

Lemma A.5 is the basis for proving the following 'iterative lemma':

A.7 Lemma (Iterative Lemma). *Suppose $Z, N \in \mathcal{A}^{\mathbb{C}}(D_\gamma(p_*, r))$, $0 \leqslant r \leqslant 1$, $0 < \delta \leqslant \gamma$ and $\epsilon \geqslant 0$ are given, with $\tilde{Z} = 0$, and assume that*

1. $$\begin{aligned} \|Z + N\|_\gamma^{p_*, r} &\leqslant 3\epsilon E \\ \|N\|_\gamma^{p_*, r} &\leqslant 2\epsilon E \end{aligned} \qquad \left(E \equiv \frac{\bar{\sigma}\bar{\rho}}{T_M}\right) \ .$$

Furthermore, assume that for some $l_1 \geqslant 1$

2. $$r \geqslant \sqrt{l_1 \epsilon}$$
3. $$\textit{and} \qquad \frac{1}{\delta^2} \frac{T_*}{T_M} r \leqslant l_1 l_2 \ ,$$

where

4. $$l_2 \equiv \min\left\{c_2, \frac{e}{2c_4 + 5c_3}\right\} \ .$$

Then there exists a symplectic diffeomorphism ϕ from $D_{\gamma-\delta}(p_*, r)$ into $D_\gamma(p_*, r)$ such that the Hamiltonian

$$\mathcal{H} \equiv H_0 + Z + N$$

is pulled back to

$$\mathcal{H} \circ \phi = H_0 + Z' + N' \ ,$$

for some $Z', N' \in \mathcal{A}^{\mathbb{C}}(D_{\gamma-\delta}(p_*, r))$ with $\tilde{Z}' = 0$ and

5. $$\|Z' + N'\|_{\gamma-\delta}^{p_*,r} \leqslant \|Z + N\|_\gamma^{p_*,r} + \|N'\|_{\gamma-\delta}^{p_*,r} \ ,$$

6. $$\|N'\|_{\gamma-\delta}^{p_*,r} \leqslant \frac{1}{e} \|N\|_\gamma^{p_*,r} \ .$$

Furthermore, for all $(g', p') \in D_{\gamma-\delta}(p_*, r)$

7. $$|p - p'| \leqslant \frac{1}{2}\delta r \bar{\rho} \qquad (\, (g, p) \equiv \phi(g', p') \,) \ .$$

The main point is that the fluctuating part N' of the transformed Hamiltonian $\mathcal{H} \circ \phi$ is less than the fluctuating part N of the untransformed Hamiltonian $\mathcal{H}$, by a factor of e. (We call N the *fluctuating part* of $\mathcal{H} = H_0 + Z + N$ because $\widetilde{H_0 + Z} = 0$.)

Note also that A.7.2 and A.7.3 are the first conditions involving the parameter r, which has been free until now (apart from the modest requirement 5.7).

PROOF. Define $\chi \equiv \mathcal{I}N \in \mathcal{A}^{\mathbb{C}}(D_\gamma(p_*, r))$, in preparation for applying Lemma A.5. Assume that δ satisfies

A.8 $$\frac{c_2 \bar{\sigma} \bar{\rho} \delta^2 r}{\|\chi\|_\gamma^{p_*,r}} \geqslant 1 \ .$$

Then Assumption B ensures that X_χ has a well-defined flow map $(t, x) \mapsto \Phi_\chi^t(x)$ as in the hypotheses of Lemma A.5, and that this map satisfies

A.9 $$\Phi_\chi^t \left(D_{\gamma-\delta}(p_*, r) \right) \subset D_{\gamma-\delta/2}(p_*, r) \qquad (-1 \leqslant t \leqslant 1) \ .$$

Furthermore, if $\phi \equiv \Phi_\chi^1$, then Assumption B also tells us that for all $(g', p') \in D_{\gamma-\delta}(p_*, r)$ the estimate A.7.7 holds.

Applying A.2.2:

A.10 $$\begin{aligned} \|\chi\|_\gamma^{p_*,r} &\leqslant \frac{1}{2} T_* \|N\|_\gamma^{p_*,r} \\ &\leqslant T_* \epsilon E = \bar{\sigma} \bar{\rho} \frac{T_*}{T_M} \epsilon \ . \end{aligned}$$

The requirement A.8 is therefore met if

$$\frac{1}{c_2} \left(\frac{\epsilon}{r^2} \right) \left(\frac{1}{\delta^2} \right) \left(\frac{T_*}{T_M} \right) r \leqslant 1 \ .$$

But this is guaranteed by the hypotheses A.7.2 and A.7.3.

For $u \in \mathcal{A}^{\mathbb{C}}(D_\gamma(p_*, r))$ one computes using the definition A.3, A.9 and Assumption C,

A.11 $$\|\Delta_1 u\|_{\gamma-\delta}^{p_*,r} \leqslant \int_0^1 \|\{u, \chi\}\|_{\gamma-\delta/2}^{p_*,r} d\tau \leqslant \left(\frac{4c_3}{\bar{\sigma}\bar{\rho}} \|u\|_\gamma^{p_*,r} \right) \frac{\|\chi\|_\gamma^{p_*,r}}{\delta^2} \ .$$

Similarly one computes from A.4, A.9 and Assumption C,

A.12
$$\|\Delta_2 u\|_{\gamma-\delta}^{p_*,r} \leqslant \left(\frac{2c_3}{\bar{\sigma}\bar{\rho}} \|u\|_{\gamma}^{p_*,r} \right) \frac{\|\chi\|_{\gamma}^{p_*,r}}{\delta^2} \ .$$

Also, by A.3, A.9 and Assumption C,

$$\begin{aligned} \|\Delta_1(H_0 - W)\|_{\gamma-\delta}^{p_*,r} &\leqslant \|\{H_0 - W, \chi\}\|_{\gamma-\delta/2}^{p_*,r} \\ &\leqslant 4c_4 \left(\sup_{p \in B_{r\bar{\rho}}^{\gamma r\bar{\rho}}(p_*)} |\Omega(p) - \Omega(p_*)| \right) \frac{\|\chi\|_{\gamma}^{p_*,r}}{\delta^2} \ . \end{aligned}$$

Since

$$\Omega(p) - \Omega(p_*) = \int_0^1 D^2 h\big(p + \tau(p_* - p)\big)(p - p_*) d\tau \ ,$$

and h is (m, M)-convex on int $B^{\bar{\rho}} \supset$ int $B_{r\bar{\rho}}^{\gamma r\bar{\rho}}(p_*)$ (see 5.1), we have

$$\sup_{p \in B_{r\bar{\rho}}^{\gamma r\bar{\rho}}(p_*)} |\Omega(p) - \Omega(p_*)| \leqslant M(\gamma r\bar{\rho} + r\bar{\rho}) \leqslant 2Mr\bar{\rho} = \frac{2r}{T_M} \ ,$$

so that we arrive at the estimate

A.13
$$\|\Delta_1(H_0 - W)\|_{\gamma-\delta}^{p_*,r} \leqslant \frac{8c_4 r}{T_M} \frac{\|\chi\|_{\gamma}^{p_*,r}}{\delta^2} \ .$$

Lemma A.5 applies, and can now use A.5.2, A.13, A.11 and A.12 to compute

$$\begin{aligned} \|N'\|_{\gamma-\delta}^{p_*,r} &\leqslant \left(\frac{8c_4 r}{T_M} + \frac{4c_3}{\bar{\sigma}\bar{\rho}} \|Z + N\|_{\gamma}^{p_*,r} + \frac{2c_3}{\bar{\sigma}\bar{\rho}} \|\tilde{N}\|_{\gamma}^{p_*,r} \right) \frac{\|\chi\|_{\gamma}^{p_*,r}}{\delta^2} \\ &\leqslant 2 \left(\frac{2c_4 r}{T_M} + \frac{5c_3 \epsilon E}{\bar{\sigma}\bar{\rho}} \right) \frac{T_* \|N\|_{\gamma}^{p_*,r}}{\delta^2} && \text{using A.7.1 and A.10} \\ &= 2(2c_4 r + 5c_3 \epsilon) \frac{1}{\delta^2} \frac{T_*}{T_M} \|N\|_{\gamma}^{p_*,r} \\ &\leqslant \left(2c_4 + \frac{5c_3}{l_1} r \right) \frac{1}{\delta^2} \frac{T_*}{T_M} r \|N\|_{\gamma}^{p_*,r} && \text{using A.7.2} \\ &\leqslant 2(2c_4 + 5c_3) \frac{1}{\delta^2} \frac{T_*}{T_M} r \|N\|_{\gamma}^{p_*,r} && \text{since } l_1 \geqslant 1 \text{ and } r \leqslant 1 \ . \end{aligned}$$

Combining this with A.7.3 gives the estimate A.7.6.

Using A.5.1, one also computes

$$\begin{aligned} \|Z' + N'\|_{\gamma-\delta}^{p_*,r} = \big\|Z + \bar{N} + N'\big\|_{\gamma-\delta}^{p_*,r} &\leqslant \big\|Z + \bar{N}\big\|_{\gamma-\delta}^{p_*,r} + \|N'\|_{\gamma-\delta}^{p_*,r} \\ &= \big\|\overline{Z + N}\big\|_{\gamma-\delta}^{p_*,r} + \|N'\|_{\gamma-\delta}^{p_*,r} \leqslant \|Z + N\|_{\gamma}^{p_*,r} + \|N'\|_{\gamma-\delta}^{p_*,r} \ , \end{aligned}$$

which proves A.7.5. □

The Hamiltonian in normal form

We now return our attention to the Hamiltonian

$$H = H_0 + F \qquad (H_0(g, p) \equiv h(p))$$

of Chap. 5, and in particular to its restriction to the domain $D_\gamma(p_*, r)$. We continue to leave γ and r as free parameters. Let $s \geqslant 1$ be given and denote by $[s] \in \mathbb{Z}$ its integer part. We seek to compose $[s]$ coordinate transformations $\phi_1, \ldots, \phi_{[s]}$ using the Iterative Lemma A.7 (with $\epsilon \equiv \|F\|/E$). At every step we will take $\delta \equiv \gamma/(2s)$ so that, assuming

$$r \geqslant \sqrt{l_1 \epsilon}$$

$$\text{and} \qquad \frac{T_*}{T_M} s^2 r \leqslant \frac{\gamma^2}{4} l_1 l_2 \ ,$$

the hypotheses A.7.2 and A.7.3 will always be satisfied.

Write $\mathcal{H}^0 \equiv H$, $Z^0 \equiv \bar{F}$ and $N^0 \equiv \tilde{F}$. A successful jth application of the Iterative Lemma will deliver a symplectic diffeomorphism ϕ^j from $D_{\gamma - j\gamma/(2s)}(p_*, r)$ into $D_{\gamma-(j-1)\gamma/(2s)}(p_*, r)$ and a new Hamiltonian $\mathcal{H}^j \equiv \mathcal{H}^{j-1} \circ \phi^j = H_0 + Z^j + N^j$, for some Z^j and N^j with $\tilde{Z}^j = 0$. In that case, writing $\alpha_j \equiv \|Z^j + N^j\|^{p_*,r}_{\gamma - j\gamma/(2s)}$ and $\beta_j \equiv \|N^j\|^{p_*,r}_{\gamma - j\gamma/(2s)}$, the inequalities A.7.5 and A.7.6 will imply

$$\text{A.14} \qquad \begin{aligned} \alpha_j &\leqslant \alpha_{j-1} + \beta_j \ , \\ \beta_j &\leqslant \frac{1}{e} \beta_{j-1} \ . \end{aligned}$$

The composition $\phi \equiv \phi_1 \circ \cdots \circ \phi_{[s]} : D_{\gamma - [s]\gamma/(2s)}(p_*, r) \to D_\gamma(p_*, r)$ will restrict to a map $\phi : D_{\gamma/2}(p_*, r) \to D_\gamma(p_*, r)$ (since $[s]/(2s) \leqslant 1/2$).

What needs to be checked is that the hypothesis A.7.1 of the Iterative Lemma holds at every step. That is, we must inductively prove that

$$\text{A.15} \qquad \begin{aligned} \alpha_{j-1} &\leqslant 3\epsilon E \ , \\ \beta_{j-1} &\leqslant 2\epsilon E \ , \end{aligned}$$

for all $1 \leqslant j \leqslant [s]$. By construction $\alpha_0 \leqslant \|F\| = \epsilon E$ and $\beta_0 \leqslant \|\tilde{F}\| \leqslant 2\|F\| = 2\epsilon E$. This anchors the induction. Suppose that A.15 (and hence A.14) holds for all $j \leqslant n$ ($n < [s]$). Then A.14 implies $\beta_n \leqslant e^{-n}\beta_0 \leqslant 2e^{-n}\epsilon E \leqslant 2\epsilon E$ and

$$\alpha_n \leqslant \alpha_0 + \left(\frac{1}{e} + \frac{1}{e^2} + \cdots + \frac{1}{e^n}\right)\beta_0 \leqslant \left(1 + \frac{2}{e-1}\right)\epsilon E \leqslant 3\epsilon E \ .$$

So A.15 also holds in the case $j = n+1$, which completes the induction.

We have $\alpha_{[s]} \leqslant \alpha_0 + (e^{-1} + \cdots + e^{-[s]})\beta_0 \leqslant 3\epsilon E$ and $\beta_{[s]} \leqslant e^{-[s]}\beta_0 \leqslant e^{-s}(2e\epsilon E)$, and summarize the preceding arguments as follows:

A.16 Proposition (The Hamiltonian in normal form). *Assume that $0 \leqslant r \leqslant 1$ and $s \geqslant 1$ satisfy*

$$r \leqslant 1 - \frac{|p_* - \bar{p}|}{\bar{\rho}} \ ,$$

$$r \geqslant \sqrt{l_1 \epsilon} \ ,$$

$$\left(\frac{T_*}{T_M}\right) s^2 r \leqslant \frac{\gamma^2}{4} l_1 l_2 \ ,$$

for some $0 \leqslant \gamma \leqslant 1$ and $l_1 > 0$ (l_2 is defined by A.7.4). Then there exists a symplectic change of coordinates $\phi : D_{\gamma/2}(p_*, r) \to D_\gamma(p_*, r)$ such that

$$H \circ \phi = H_0 + Z + N \ ,$$

for some $Z, N \in \mathcal{A}^{\mathbb{C}}(D_{\gamma/2}(p_*, r))$, with $\tilde{Z} = 0$ and

$$\|Z + N\|^{p_*,r}_{\gamma/2} \leqslant 3\epsilon E \ ,$$

$$\|N\|^{p_*,r}_{\gamma/2} \leqslant e^{-s}(2e\epsilon E) \ .$$

Furthermore, for all $(g', p') \in D_{\gamma/2}(p_*, r)$

$$|p - p'| \leqslant \frac{\gamma}{4} r\bar{\rho} \qquad (\,(g,p) \equiv \phi(g',p')\,) \ .$$

The first of the three hypotheses is just 5.7, repeated here for completeness.

Moving coordinate systems and reduction to the zero frequency case

To expose the role played by convexity[1] in establishing bounds on the perturbed dynamics, we prefer over Lochak's argument one based on moving coordinate transformations that we describe next.

Proposition A.16 says that there exists a change of coordinates ϕ such that $H \circ \phi = H_0 + Z + N$, for some Z and N with $\tilde{Z} = 0$ and such that N is exponentially small with respect to the parameter s. Restrict attention to the real domain $D_0(p_*, r) = G \times B_{r\bar{\rho}}(p_*)$, and suppose that $\Omega(p_*) = 0$ ($\Omega \equiv \nabla h$, $H_0(g,p) = h(p)$). Then the convexity of h and the Morse lemma imply that in a neighborhood of $G \times \{p_*\}$ the level sets of H_0 are diffeomorphic to $G \times S^k$, where S^k denotes the unit sphere in $\mathfrak{t}$, centered at p_*. In particular these level sets are *compact.* Now $\|Z + N\|_0^{p_*,r} \leqslant 3\epsilon E$ (by A.16), so for ϵ sufficiently small the level sets of $H \circ \phi$ are also compact (in some neighborhood of $G \times \{p_*\}$) and energy conservation implies that integral curves of $X_{H\circ\phi}$, beginning sufficiently close to $G \times \{p_*\}$, remain $\sqrt{\epsilon}$-close *for all time.*

One attempts to convert the case $\Omega(p_*) \neq 0$ to the zero frequency case just described by making an appropriate moving coordinate transformation. The following lemma shows how to compute solution curves in a Hamiltonian system, as they appear in a moving coordinate system, by computing solutions to an associated *nonautonomous* system. The proof of the lemma is elementary and left to the reader.

A.17 Lemma (On moving coordinate transformations). *Let $\mathcal{H}$ and W be differentiable functions on $D_0(p_*, r)$ and assume X_W has a well-defined flow map $(t,x) \mapsto \Phi_W^t(x) : \mathbb{R} \times D_0(p_*, r) \to D_0(p_*, r)$. Suppose $t \mapsto x_t \in D_0(p_*, r)$ is an integral curve of $X_{\mathcal{H}}$ and define $x'_t \equiv (\Phi_W^t)^{-1}(x_t)$ (i.e., x'_t is the point x_t as seen by the 'moving coordinates' $(\Phi_W^t)_{t\in\mathbb{R}}$). Then $t \mapsto x'_t$ is an integral curve of the nonautonomous system $(X_{\mathcal{H}'_t})_{t\in\mathbb{R}}$, where $\mathcal{H}'_t$ is the time-dependent Hamiltonian defined by*

$$\mathcal{H}'_t \equiv \mathcal{H} \circ \Phi_W^t - W \ .$$

By $t \mapsto x'_t$ being an integral curve of the nonautonomous system $(X_{\mathcal{H}'_t})_{t\in\mathbb{R}}$, we mean that

$$\frac{d}{dt} x'_t = X_{\mathcal{H}'_t}(x'_t) \qquad (t \in \mathbb{R}) \ .$$

Taking $W(g,p) \equiv w(p) = \Omega(p_*) \cdot p$ and $\mathcal{H} = H \circ \phi = H_0 + Z + N$ (with $\Omega(p_*)$ possibly non-zero), we have

$$\mathcal{H}'_t = (H_0 - W) + Z + N \circ \Phi_W^t$$

(because H_0 and Z are Φ_W^t-invariant). Or, defining $H'_0 \equiv H_0 - W$ and $N_t \equiv N \circ \Phi_W^t$, we may write

$$\mathcal{H}'_t = H'_0 + Z + N_t \ .$$

---

[1] Here we are thinking of the *first* estimate in 5.3, which we have yet to employ.

We can write $H_0'(g,p) \equiv h'(p)$ if we define $h' \equiv h - w$. Then if $\Omega' \equiv \nabla h'$, we have $\Omega'(p_*) = 0$. We are therefore in the zero frequency situation already described, with the important exception that we are dealing with a *time-dependent* perturbation $Z + N_t$.

Time-dependence destroys energy conservation, and therefore stability, as we argued above. However, the smaller the magnitude of the time-dependence (i.e., of $\partial N_t/\partial t$), the closer to stability one expects to get. Indeed this is the case:

A.18 Proposition. *Let $Z, N_t : D_0(p_*, r) \to \mathbb{R}$ be differentiable functions ($t \in \mathbb{R}$) such that $\tilde{Z} = 0$ and such that $t \mapsto N_t(x)$ is differentiable for all x in the interior of $D_0(p_*, r)$. Let $\alpha \geqslant 0$ be any number such that*

$$\|Z + N_0\|_0^{p_*,r} \leqslant \alpha E \ ,$$

and assume that for some positive $r_1 \leqslant r$ one has

$$1. \qquad r_1^2 \geqslant \frac{16\bar{\sigma}\alpha}{c_7} \ .$$

Consider the time-dependent Hamiltonian

$$\mathcal{H}_t \equiv H_0' + Z + N_t \ , \qquad (t \in \mathbb{R})$$

where $H_0' \equiv H_0 - W$, and an integral curve $t \mapsto (g_t, p_t)$ of the nonautonomous system $(X_{\mathcal{H}_t})_{t\in\mathbb{R}}$. Then one has

$$\left(\frac{|p_0 - p_*|}{\bar{\rho}} \leqslant c_7' r_1 \quad \text{and} \quad |t| \leqslant t_0 \right) \implies \frac{|p_t - p_*|}{\bar{\rho}} \leqslant r_1 \ ,$$

where

$$c_7' \equiv \frac{\sqrt{5}-2}{2} c_7 \qquad \text{and} \qquad t_0 \equiv \frac{\alpha}{\left(\sup_{t\in\mathbb{R}} \|\frac{\partial N_t}{\partial t}\|_0^{p_*,r}\right)/E} \ .$$

Note that $0 < c_7' < 1$, since $c_7 = m/M \leqslant 1$. Notice also of course that the 'stability time' t_0 approaches ∞ as $\|\partial N_t/\partial t\|_0^{p_*,r} \to 0$.

Proof. Suppose then that $|p_0 - p_*| \leqslant c_7' \bar{\rho} r_1$, with c_7' as above. Then by the compactness of $D_0(p_*, c_7' r_1) = G \times B_{c_7'\bar{\rho}r_1}(p_*)$, the curve $t \mapsto (g_t, p_t) \in D_0(p_*, r_1)$ is well-defined until it reaches the boundary of $D_0(p_*, r_1)$. That is, there exists a time st_0 ($s = \pm 1, 0 < t_0 \leqslant \infty$) such that

$$\begin{aligned} &|p_{st_0} - p_*| = \bar{\rho} r_1 \\ \text{and} \quad &|t| \leqslant t_0 \implies (g_t, p_t) \in D_0(p_*, r_1) \ . \end{aligned}$$

Write $x_t \equiv (g_t, p_t)$. Then the analogue of energy conservation for nonautonomous Hamiltonians is the identity

$$\text{A.19} \qquad \frac{d}{dt}\left(\mathcal{H}_t(x_t)\right) = \frac{\partial \mathcal{H}_t}{\partial t}(x_t) \ .$$

Recall that one proves this by applying the chain rule and computing

$$\frac{d}{d\tau}\mathcal{H}_t(x_\tau)|_{\tau=t} = \langle d\mathcal{H}_t, \frac{d}{dt}x_t \rangle = \omega(X_{\mathcal{H}_t}(x_t), X_{\mathcal{H}_t}(x_t)) = 0 \ .$$

Substituting $\mathcal{H}_t = H_0' + Z + N_t$ into A.19 and integrating leads to

$$h'(p_{st_0}) - h'(p_0) = \int_0^{st_0} \frac{\partial N_\tau}{\partial \tau}(x_\tau)d\tau - \big((Z + N_{st_0})(x_{st_0}) - (Z + N_0)(x_0) \big) ,$$

where $h' \equiv h - w$. Since $D^2h' = D^2h$, we can apply Taylor's formula with integral remainder to conclude

$$\text{A.20} \quad \int_0^1 (1-\tau)D^2h(p_0 + \tau\delta p)(\delta p, \delta p)d\tau = \int_0^{st_0} \frac{\partial N_\tau}{\partial \tau}(x_\tau)d\tau - \big((Z + N_{st_0})(x_{st_0}) - (Z + N_0)(x_0) \big) - Dh'(p_0)\delta p ,$$

where $\delta p \equiv p_{st_0} - p_0$. We will arrive at a lower bound on t_0 proving our proposition by bounding the LHS of A.20 from below and the terms on the RHS from above. Indeed, by the (m, M)-convexity of h, we have

$$\text{A.21} \quad \int_0^1 (1-\tau)D^2h(p_0 + \tau\delta p)(\delta p, \delta p)d\tau \geqslant \frac{1}{2}m|\delta p|^2 = \frac{c_7 E|\delta p|^2}{2\bar{\rho}^2\bar{\sigma}} .$$

Furthermore, we have

$$\text{A.22} \quad \left| \int_0^{st_0} \frac{\partial N_\tau}{\partial \tau}(x_\tau)d\tau \right| \leqslant t_0 \sup_{t\in\mathbb{R}} \left\| \frac{\partial N_t}{\partial t} \right\|_0^{p_*, r} ,$$

and

$$\begin{aligned} |(Z + N_{st_0})(x_{st_0}) - (Z + N_0)(x_0)| &\leqslant |(Z + N_0)(x_{st_0}) - (Z + N_0)(x_0)| \\ &\quad + |N_{st_0}(x_{st_0}) - N_0(x_{st_0})| \\ &\leqslant 2\,\|Z + N_0\|_0^{p_*, r} + \left| \int_0^{st_0} \frac{\partial N_\tau}{\partial \tau}(x_{st_0})d\tau \right| \\ \text{A.23} \quad &\leqslant 2\alpha E + t_0 \sup_{t\in\mathbb{R}} \left\| \frac{\partial N_t}{\partial t} \right\|_0^{p_*, r} . \end{aligned}$$

Finally, as $D^2h = D^2h'$ and $Dh'(p_*) = \Omega(p_*) - \Omega(p_*) = 0$, we have

$$\begin{aligned} |Dh'(p_0)\delta p| &= \left| \int_0^1 Dh^2(p_* + \tau(p_0 - p_*))(p_0 - p_*, \delta p)d\tau \right| \\ \text{A.24} \quad &\leqslant Mc_7'\bar{\rho}r_1|\delta p| = \frac{c_7' E r_1 |\delta p|}{\bar{\rho}\bar{\sigma}} . \end{aligned}$$

Applying the estimates A.21–A.24 to A.20, we obtain

$$\frac{c_7 E|\delta p|^2}{2\bar{\rho}^2\bar{\sigma}} \leqslant 2\alpha E + 2t_0 \sup_{t\in\mathbb{R}} \left\| \frac{\partial N_t}{\partial t} \right\|_0^{p_*, r} + \frac{c_7' r_1 E|\delta p|}{\bar{\rho}\bar{\sigma}} .$$

Since $\delta p = p_{st_0} - p_0$, $|p_{st_0} - p_*| = \bar{\rho}r_1$ and $|p_0 - p_*| \leqslant c_7'\bar{\rho}r_1$, we have

$$(1 - c_7')\bar{\rho}r_1 \leqslant |\delta p| \leqslant (1 + c_7')\bar{\rho}r_1 ,$$

so that the preceding estimate yields

$$t_0 \geqslant \frac{E}{\sup_{t\in\mathbb{R}} \left\| \frac{\partial N_t}{\partial t} \right\|_0^{p_*, r}} \left(\left(\frac{c_7(1 - c_7')^2}{4} - \frac{c_7'(1 + c_7')}{2} \right) \frac{r_1^2}{\bar{\sigma}} - \alpha \right) .$$

The reader can check (recalling that $c_7' \equiv (\sqrt{5}-2)c_7/2$ and $0 < c_7 \leqslant 1$) that the coefficient of $r_1^2/\bar{\sigma}$ is bounded from below by $c_7/8$. Together with our hypothesis on r_1^2, this yields

$$t_0 \geqslant \frac{\alpha}{\left(\sup_{t\in\mathbb{R}} \left\|\frac{\partial N_t}{\partial t}\right\|_0^{p_*,r}\right)/E} .$$

□

Nekhoroshev estimates in the neighborhood of the periodic orbits

We now combine our preceding arguments to deduce Nekhoroshev-type estimates on motions in a neighborhood of $G \times \{p_*\}$, which by assumption is foliated by periodic orbits of X_H.

Let $r \geqslant 0$ be given and define $R \equiv 4r/c_7' \geqslant 4r$ ($c_7' \in (0,1)$ being the constant defined in A.18). Assume

$$R \leqslant 1 - \frac{|p_* - \bar{p}|}{\bar{\rho}} ,$$

A.25
$$R \geqslant \sqrt{l_1\epsilon} ,$$

$$\frac{T_*}{T_M} s^2 R \leqslant \frac{\gamma^2}{4} l_1 l_2 ,$$

for some $l_1 > 0$. Then Proposition A.16 applies, with r replaced with R, delivering some symplectic coordinate change $\phi : D_{\gamma/2}(p_*, R) \to D_\gamma(p_*, R)$.

Let $({}^1p_0, {}^1g_0) \in D_0(p_*, r) \subset D_0(p_*, R)$ be given. Then

$$|{}^1p_0 - p_*| \leqslant \bar{\rho} r .$$

For t in some open interval of zero, there exists an integral curve $t \mapsto ({}^1g_t, {}^1p_t) \in D_0(p_*, R)$ of X_H with initial condition $({}^1g_0, {}^1p_0)$ (remember $r \leqslant R/4 < R$). Our objective is to derive an estimate on the evolution of ${}^1p_t$.

Define

$$({}^2g_t, {}^2p_t) \equiv \phi^{-1}({}^1g_t, {}^1p_t) .$$

Then $t \mapsto ({}^2g_t, {}^2p_t)$ is an integral curve of $X_{H\circ\phi}$. According to the last estimate in Proposition A.16,

$$|{}^2p_0 - {}^1p_0| \leqslant \frac{\gamma}{4} \bar{\rho} R ,$$

from which we deduce

$$|{}^2p_0 - p_*| \leqslant \bar{\rho}\left(\frac{\gamma}{4}R + r\right) = \bar{\rho} r \left(\frac{\gamma}{c_7'} + 1\right) .$$

Choose $\gamma \equiv c_7'$. Then

$$|{}^2p_0 - p_*| \leqslant 2\bar{\rho} r .$$

Define

$$({}^3g_t, {}^3p_t) \equiv (\Phi_W^t)^{-1}({}^2g_t, {}^2p_t) .$$

By properties of Φ_W^t established at the beginning of this appendix, we known that ${}^3p_0 = {}^2p_0$, so that the above estimate implies

$$|{}^3p_0 - p_*| \leqslant 2\bar{\rho} r .$$

By Lemma A.17 and the arguments that followed, $t \mapsto ({}^3g_t, {}^3p_t)$ is a solution curve for the nonautonomous Hamiltonian

$$\mathcal{H}_t \equiv H_0' + Z + N_t \ ,$$

where $H_0' \equiv H_0 - W$, $N_t \equiv N \circ \Phi_W^t$ and where $Z, N \in D_{\gamma/2}(p_*, R) = D_{c_7'/2}(p_*, R)$ are the functions delivered by Proposition A.16 (with r replaced with R and $\gamma \equiv c_7'$). In particular, we know from A.16 that

$$\|Z + N\|_{c_7'/2}^{p_*,R} \leqslant 3\epsilon E \ ,$$
$$\|N\|_{c_7'/2}^{p_*,R} \leqslant e^{-s}(2e\epsilon E) \ .$$

Since $N_0 = N$ and $\|N_t\|_{c_7'/2}^{p_*,R} = \|N \circ \Phi_W^t\|_{c_7'/2}^{p_*,R} = \|N\|_{c_7'/2}^{p_*,R}$ (by forementioned properties of Φ_W^t), we deduce

$$\begin{aligned} \|Z + N_0\|_0^{p_*,R} &\leqslant 3\epsilon E \ , \\ \left\|\frac{\partial N_t}{\partial t}\right\|_0^{p_*,R} &= \|\{N_t, W\}\|_0^{p_*,R} \\ &\leqslant \frac{4c_4}{{c_7'}^2}|\Omega(p_*)| \, \|N_t\|_{c_7'/2}^{p_*,R} \qquad \text{by Assumption C} \\ &\leqslant e^{-s}\left(\frac{8ec_4}{{c_7'}^2 T_\Omega}\epsilon E\right) \ . \end{aligned}$$

We now apply Proposition A.18 with $\alpha \equiv 3\epsilon$, $r_1 \equiv 2r/c_7' = R/2 < R$, and with r in A.18 replaced by R. To satisfy A.18.1 we require $r \geqslant \sqrt{12{c_7'}^2\bar{\sigma}\epsilon/c_7}$. The condition A.25 already guarantees this if the free parameter $l_1 \geqslant 1$ is subject to the additional condition

$$l_1 \geqslant 192\bar{\sigma}/c_7 \ .$$

Since $|{}^3p_0 - p_*| \leqslant 2\bar{\rho}r = \bar{\rho}c_7' r_1$, we can apply A.18 to the trajectory $t \mapsto ({}^3g_t, {}^3p_t)$ to deduce

$$|t| \leqslant t_0 \implies \frac{|{}^3p_t - p_*|}{\bar{\rho}} \leqslant \frac{2r}{c_7'} \ ,$$

where

$$t_0 \equiv \frac{\alpha}{\left(\sup_{t\in\mathbb{R}} \left\|\frac{\partial N_t}{\partial t}\right\|_0^{p_*,R}\right)/E} \geqslant \frac{3{c_7'}^2 T_\Omega}{8ec_4} e^s \ .$$

Now ${}^3p_t = {}^2p_t$, and by the last estimate in Proposition A.16 (with r replaced with R and $\gamma \equiv c_7'$),

$$|{}^2p_t - {}^1p_t| \leqslant \frac{c_7'}{4} R\bar{\rho} = \bar{\rho}r \ .$$

In particular,

$$|{}^1p_t - p_*| \leqslant |{}^3p_t - p_*| + \bar{\rho}r \ ,$$

so that

$$|t| \leqslant t_0 \implies \frac{|{}^1p_t - p_*|}{\bar{\rho}} \leqslant \left(1 + \frac{2}{c_7'}\right) r \ .$$

Note that $(1+2/c_7')r \leqslant R/4 + R/2 < R$.

Summarizing:

A.26 Theorem (c.f. Theorem 1, Lochak (1993)). *Consider the Hamiltonian $H = H_0 + F$ considered in Chap. 5 ($H_0(g,p) = h(p)$). Assume that $\Omega(p_*) = \mathbf{n}/T_*$, for some $p_* \in \operatorname{int} B$, $\mathbf{n} \in \mathfrak{t}^{\mathbb{Z}}$ and $T_* > 0$ ($\Omega \equiv \nabla h$). Furthermore assume that for some $r \geqslant 0$, $s \geqslant 1$ and $l_1 \geqslant \max\{1, 192\bar{\sigma}/c_7\}$ we have*

$$r \leqslant \frac{c_7'}{4}\left(1 - \frac{|p_* - \bar{p}|}{\bar{\rho}}\right) ,$$

$$r \geqslant \frac{c_7'}{4}\sqrt{l_1 \epsilon} ,$$

$$\frac{T_*}{T_M} s^2 r \leqslant \frac{{c_7'}^3}{16} l_1 l_2 ,$$

where $c_7' \equiv (\sqrt{5}-2)c_7/2$ and l_2 is given by A.7.4. Then (real) integral curves $t \mapsto (g_t, p_t) \in G \times B$ of X_H satisfy the estimate

$$\left(\frac{|p_0 - p_*|}{\bar{\rho}} \leqslant r \quad \text{and} \quad \frac{|t|}{T_\Omega} \leqslant \frac{3{c_7'}^2}{8ec_4} e^s\right) \Longrightarrow \frac{|p_t - p_*|}{\bar{\rho}} \leqslant \left(1 + \frac{2}{c_7'}\right) r .$$

Taking $s \equiv \epsilon^{-b/2}$ and $r \equiv ({c_7'}^3/16) l_1 l_2 (T_M/T_*) \epsilon^b$ for some $b > 0$:

A.27 Corollary (c.f. Corollary 3, Lochak (1993)). *Suppose $l_1 \geqslant \max\{1, 192\bar{\sigma}/c_7\}$ and $b > 0$ are given such that*

$$\frac{{c_7'}^2 l_1 l_2}{A_*} \epsilon^b \leqslant \frac{T_*}{T_M} \leqslant \frac{{c_7'}^2 l_1^{1/2} l_2}{4} \epsilon^{b-1/2} ,$$

where

$$A_* \equiv \left(1 - \frac{|p_* - \bar{p}|}{\bar{\rho}}\right)$$

and l_2 is given by A.7.4. Then

$$\left(\frac{|p_0 - p_*|}{\bar{\rho}} \leqslant \frac{{c_7'}^3 l_1 l_2}{16} \frac{\epsilon^b}{T_*/T_M} \quad \text{and} \quad \frac{|t|}{T_\Omega} \leqslant \frac{3{c_7'}^2}{8ec_4} \exp(\epsilon^{-b/2})\right) \Longrightarrow$$

$$\frac{|p_t - p_*|}{\bar{\rho}} \leqslant \left(1 + \frac{2}{c_7'}\right) \frac{{c_7'}^3 l_1 l_2}{16} \frac{\epsilon^b}{T_*/T_M} .$$

Estimates for an arbitrary initial condition

We now seek an estimate on solutions $t \mapsto (g_t, p_t)$ with $p_0 \equiv \bar{p}$. Recall that $\bar{p}$ is a point in $\mathfrak{t}_0$, subject to the condition that the Hamiltonian is defined in some complex neighborhood $G^{\bar{\sigma}} \times B^{\bar{\rho}}$ of $G \times \{\bar{p}\}$.

Lochak's marvelous observation is that a stability time for any initial condition can be deduced from those of the periodic orbits studied above. To do so requires understanding how the periodic orbits are distributed, and the relevant tool is purely arithmetic.

View $\{\beta_1, \ldots, \beta_k\}$ (the $\mathbb{Z}$-basis for $\mathfrak{t}^{\mathbb{Z}}$) as an $\mathbb{R}$-basis for $\mathfrak{t}$, and let $\gamma : \mathbb{R}^k \to \mathfrak{t}$ denote the associated isomorphism. The l^∞-norm on $\mathbb{R}^k$ induces a norm $|\cdot|_\infty$ on $\mathfrak{t}$:

$$|\gamma(\tau_1, \ldots, \tau_k)|_\infty \equiv \max_{1 \leqslant j \leqslant k} |\tau_j| \qquad (\tau_j \in \mathbb{R}) .$$

We have $|\xi| \leqslant c_1\sqrt{k}|\xi|_\infty$ ($\xi \in \mathfrak{t}$), courtesy of 5.6.

Since $\gamma(\mathbb{Z}^k) = \mathfrak{t}^{\mathbb{Z}}$, Dirichlet's simultaneous approximation theorem (see, e.g., Schmidt (1980)) tells us that

$$(\forall \varpi \in \mathfrak{t})(\forall Q > 1)(\exists q \in [1, Q) \cap \mathbb{Z}) : \ |q\varpi - \mathfrak{t}^{\mathbb{Z}}|_\infty \leqslant \frac{1}{Q^{1/k}} \ .$$

Taking $\varpi \equiv \Omega(\bar{p})$ and writing $T_* \equiv T_M q$, one deduces:

A.28 Theorem (On the proximity of periodic orbits).
For all $Q > 1$ there exists $T_* > 0$ and $\Omega_* \in \frac{1}{T_*}\mathfrak{t}^{\mathbb{Z}}$ such that

$$1 < \frac{T_*}{T_M} \leqslant Q \quad \text{and} \quad |\Omega(\bar{p}) - \Omega_*| \leqslant \frac{c_1\sqrt{k}}{T_* Q^{1/k}} \ .$$

Because of the convexity of h, the frequency map is locally invertible:

A.29 Lemma. *Define $c_5' \equiv \min\{c_5/2, 1\}$. The map $\Omega : B \to \mathfrak{t}$ has a locally defined inverse Ω^{-1} mapping $B_{c_7 c_5'/(2T_M)}(\Omega(\bar{p}))$ diffeomorphically onto an open subset of $B = B_{\bar{\rho}}(\bar{p})$.*

PROOF. Apply the inverse function theorem, as formulated in, e.g., Proposition 2.5.6 of Abraham et al. (1988)[2]. □

If $\Omega_* \in B_{c_7 c_5'/(2T_M)}(\Omega(\bar{p}))$ and $p* \equiv \Omega^{-1}(\Omega_*)$, then one computes

$$\begin{aligned} |\bar{p} - p_*| &= |\Omega^{-1}(\Omega(\bar{p})) - \Omega^{-1}(\Omega_*)| \\ &= \left| \int_0^1 D\Omega^{-1}(\Omega_* + t\delta\Omega)\delta\Omega dt \right| \qquad (\,\delta\Omega \equiv \Omega(\bar{p}) - \Omega_*\,) \\ &= \left| \int_0^1 [D^2 h(\Omega^{-1}(\Omega_* + t\delta\Omega))]^{-1} \delta\Omega dt \right| \\ &\leqslant \frac{1}{m}|\delta\Omega| = \bar{\rho} T_m |\Omega(\bar{p}) - \Omega_*| \ . \end{aligned}$$

With the aid of this estimate, the reader may readily verify that A.28 and A.29 combine to yield:

A.30 Corollary. *Assume*

$$Q^{1/k} \geqslant \max\{\frac{2c_1\sqrt{k}}{c_7 c_5'}, 1\} \ .$$

Then there exists $T_* > 0$ and $p_* \in B$ with $\Omega(p_*) \in \frac{1}{T_*}\mathfrak{t}^{\mathbb{Z}}$ such that

$$1 < \frac{T_*}{T_M} \leqslant Q \quad \text{and} \quad \frac{|\bar{p} - p_*|}{\bar{\rho}} \leqslant \frac{c_1\sqrt{k}}{c_7 Q^{1/k}(T_*/T_M)} \ .$$

Let $b > 0$ and $l_1 \geqslant \max\{1, 192\bar{\sigma}/c_7\}$ be given. If we choose

$$Q^{1/k} \equiv \frac{16 c_1 \sqrt{k}}{c_7 {c_7'}^3 l_1 l_2} \epsilon^{-b} \ ,$$

[2]Their statement contains a typographical error. Correct values for the constants given are $P \equiv \min\{(2KM)^{-1}, R\}$, $Q \equiv \min\{(2NL)^{-1}, P/(2M)\}$, $S \equiv \min\{(2KM)^{-1}, Q/(2L)\}$.

then the hypothesis of A.30 is satisfied if

$$\epsilon \leqslant \left(\frac{c_7 {c'_7}^3 l_1 l_2}{8 l_3} \right)^{1/b} ,$$

where $l_3 \equiv \min\{c_7 c'_5, 2c_1\sqrt{k}\}$. In that case, for some $T_* > 0$, with

$$\text{A.31} \qquad 1 < \frac{T_*}{T_M} \leqslant \left(\frac{16 c_1 \sqrt{k}}{c_7 {c'_7}^3 l_1 l_2} \right)^k \epsilon^{-kb} ,$$

there exists $p_* \in B$ such that $\Omega(p_*) \in \frac{1}{T_*}\mathfrak{t}^{\mathbb{Z}}$ and

$$\text{A.32} \qquad \frac{|\bar{p} - p_*|}{\bar{\rho}} \leqslant \frac{{c'_7}^3 l_1 l_2}{16} \frac{\epsilon^b}{T_*/T_M} .$$

If we can satisfy the hypothesis on T_*/T_M in Corollary A.27, then for any solution $t \mapsto (g_t, p_t)$ with $p_0 = \bar{p}$ we will be able to conclude from A.32 (using $|p_t - \bar{p}| \leqslant |p_t - p_*| + |\bar{p} - p_*|$ and $T_*/T_M > 1$) that

$$\text{A.33} \qquad \frac{|t|}{T_\Omega} \leqslant \frac{3{c'_7}^2}{8ec_4} \exp(\epsilon^{-b/2}) \implies \frac{|p_t - \bar{p}|}{\bar{\rho}} \leqslant \left(2 + \frac{2}{c'_7}\right) \frac{{c'_7}^3 l_1 l_2}{16} \epsilon^b .$$

By A.31 the hypothesis on T_*/T_M in A.27 is met provided

$$\text{A.34} \qquad \epsilon \leqslant \left(\frac{A_*}{{c'_7}^2 l_1 l_2} \right)^{1/b}$$

$$\text{A.35} \qquad \text{and} \quad \epsilon^{1/2-(1+k)b} \leqslant l_4 l_1^{k+1/2} ,$$

where

$$l_4 \equiv \frac{{c'_7}^2 l_2}{4} \left(\frac{c_7 {c'_7}^3 l_2}{16 c_1 \sqrt{k}} \right)^k .$$

We observe from A.31, A.32 and the definition of A_*, that

$$A_* \geqslant 1 - \frac{{c'_7}^3 l_1 l_2}{16} \epsilon^b .$$

So we ensure $A_* \geqslant 1/2$ if we assume

$$\epsilon \leqslant \left(\frac{8}{{c'_7}^3 l_1 l_2} \right)^{1/b} .$$

The inequality A.34 is guaranteed in that case provided

$$\epsilon \leqslant \left(\frac{1}{2{c'_7}^2 l_1 l_2} \right)^{1/b} .$$

What remains is to ensure A.35. Remember that l_1 is still a free parameter, apart from the requirement $l_1 \geqslant \max\{1, 192\bar{\sigma}/c_7\}$. Choose

$$l_1 \equiv \max \left\{ 1, \frac{192\bar{\sigma}}{c_7}, \left(\frac{1}{l_4} \right)^{k+1/2} \right\} .$$

Then to satisfy the condition A.35 it suffices to ensure $\epsilon^{1/2-(k+1)b} \leqslant 1$, which we achieve with the choice

$$b = \frac{1}{2(1+k)} \ .$$

To summarize, the exponential estimate A.33 holds, with l_1 and b as above, provided

$$\epsilon_0 = \min \left\{ \left(\frac{8}{c_7'^3 l_1 l_2} \right)^{1/b}, \left(\frac{1}{2c_7'^2 l_1 l_2} \right)^{1/b} \right\} \ .$$

This completes the proof of Theorem 5.8 and its Addendum 5.9.

APPENDIX B

Proof that $\mathcal{W}$ is a Slice

Suppose G acts smoothly on a smooth manifold M. Recall that S is a *slice* at $x \in M$ if S is a G_x-invariant submanifold of M ($G_x \subset G$ the isotropy subgroup at x) containing x and such that

$$G \times_{G_x} S \to M \ , \qquad [g, x] \mapsto g \cdot s$$

is a diffeomorphism onto some open neighborhood of $G \cdot x$. Here $G \times_{G_x} S$ is the quotient $(G \times S)/G_x$, where G_x acts on $G \times S$ according to $h \cdot (g, s) \equiv (gh^{-1}, h \cdot x)$, and $[g, s] \equiv (g, s) \mod G_x$. We call S a *global* slice if the image of the map above is M. Sufficient conditions for a G_x-invariant submanifold $S \ni x$ to be a slice at x are: (i) S intersects G-orbits in a 'complementary fashion,' i.e., $\mathrm{T}_s M = \mathrm{T}_s S \oplus \mathrm{T}_s(G \cdot s)$ ($s \in S$), and (ii) $s \in S$, $g \in G$ and $g \cdot s \in S$ together imply $g \in G_x$.

Let us now verify that the (open) Weyl chamber $\mathcal{W}$ in $\mathfrak{g}^*$ is a global slice for the co-adjoint action of G on $\mathfrak{g}^*_{\mathrm{reg}}$. Note first that $\mathcal{W}$ is G_w-invariant for any $w \in \mathcal{W}$, by Corollary 1.8.2, Part 1. Since each regular co-adjoint orbit intersects $\mathcal{W}$ in a unique point (see, e.g., Corollary 1.8.3, Part 1), it follows that $G(\mathcal{W}) = \mathfrak{g}^*_{\mathrm{reg}}$ (so that if $\mathcal{W}$ is indeed a slice, then it is a global slice) and

$$(w \in \mathcal{W} \ , g \in G \ , \text{ and } g \cdot w \in \mathcal{W}) \Rightarrow g = \mathrm{id} \ .$$

By the above, it remains only to show that $\mathcal{W}$ intersects co-adjoint orbits in a complementary fashion, i.e.,

B.1 $$\mathrm{T}_w \mathfrak{g}^* = \mathrm{T}_w(G \cdot w) \oplus \mathrm{T}_w \mathcal{W} \qquad (w \in \mathcal{W}) \ .$$

We claim, identifying the various spaces with subspaces of $\mathfrak{g}^*$, that

B.2 $$\mathrm{T}_w(G \cdot w) = \underline{\mathfrak{t}}^{\perp} \equiv \operatorname{Ann} \mathfrak{t}$$

B.3 $$\mathrm{T}_w \mathcal{W} = \underline{\mathfrak{t}} \equiv \operatorname{Ann}[\mathfrak{g}, \mathfrak{t}] \ ,$$

for all $w \in \mathcal{W}$. Then B.1 follows immediately from the direct sum decomposition

$$\mathfrak{g}^* = \underline{\mathfrak{t}}^{\perp} \oplus \underline{\mathfrak{t}} \ ;$$

see, e.g., Sect. 1, Part 1.

Since $\mathcal{W} \subset \underline{\mathfrak{t}}$ is open, B.3 is obvious.

Proof of B.2. Since regular co-adjoint orbits are of type (T) (by, e.g., 1.8.1, 1.8.2 and 1.2.2 of Part 1), we know that $\dim \mathrm{T}_w(G \cdot w) = \dim \underline{\mathfrak{t}}$ ($w \in \mathcal{W} \subset \underline{\mathfrak{t}} \cap \mathfrak{g}^*_{\mathrm{reg}}$). To prove B.2 it therefore suffices to show that $\mathrm{T}_w(G \cdot w) \subset \underline{\mathfrak{t}}^{\perp} = \operatorname{Ann} \mathfrak{t}$. Elements β of $\mathrm{T}_w(G \cdot w)$ are of the form $\beta = \mathrm{ad}^*_\xi w$ for some $\xi \in \mathfrak{g}$. But then for all $\tau \in \mathfrak{t}$ we compute

$$\langle \beta, \tau \rangle = \langle \mathrm{ad}^*_\xi w, \tau \rangle = -\langle \mathrm{ad}^*_\tau w, \xi \rangle = 0 \ ,$$

since $\mathfrak{g}_w = \mathfrak{t}$ (by, e.g., Corollary 1.8, Part 1). This shows that $\beta \in \operatorname{Ann} \mathfrak{t} = \underline{\mathfrak{t}}^{\perp}$. Since $\beta \in \mathrm{T}_w(G \cdot w)$ was arbitrary, this proves $\mathrm{T}_w(G \cdot w) \subset \underline{\mathfrak{t}}^{\perp}$, as required. $\square$

APPENDIX C

Proof of the Extension Lemma

This appendix is devoted to a proof of Lemma 8.8. Recall the fact (following from the last statement in Theorem 8.8) that every point of P is of the form $g \cdot y$ for some $g \in G$ and $y \in F$. We shall use this fact repeatedly in the sequel without further comment.

Let us check that the extended action of H is well-defined. As in the proof of Theorem 8.14, we have $g' \cdot y' = g \cdot y$ ($g, g' \in G$; $y, y' \in F$) if and only if $g' = gq$ and $y' = q^{-1} \cdot y$ for some $q \in T$. In that case, for any $h \in H$, we obtain $g' \cdot (h \odot y') = (gq) \cdot (h \odot (q^{-1} \cdot y)) = g \cdot (h \odot y)$, since the actions of T and H on F commute by hypothesis. This shows that the extended action is well-defined.

That the extension of $\mathbf{K}$ is well-defined follows similarly, appealing to hypothesis 2.

Next, we argue that the extended action of H is symplectic. Fix some $h \in H$ and define $\phi : P \to P$ by $\phi(x) \equiv h \odot x$. We need to show that ϕ is symplectic. Since G acts symplectically on P, and H acts symplectically on F, it follows from the definition of the extended action that ϕ maps fibers of $\pi : P \to \mathcal{O}$ symplectically onto themselves. It therefore remains only to show that $\mathrm{T}\phi$ maps horizontal spaces of the symplectic connection on $\pi : P \to \mathcal{O}$ symplectically onto horizontal spaces. From the definition of ϕ, the extended action of H, and infinitesimal generators, it follows that

$$\text{C.1} \qquad \mathrm{T}\phi \cdot \xi_P(g \cdot y) = \xi_P(\phi(g \cdot y)) \qquad (\xi \in \mathfrak{g},\, g \in G,\, y \in F)\ .$$

Since $\mathbf{J}$ is G-equivariant and $\mathbf{J}|F$ is H-invariant (hypothesis 3), we compute

$$\begin{aligned}\text{C.2} \quad \mathbf{J}(\phi(g \cdot y)) &= \mathbf{J}(h \odot (g \cdot y)) = \mathbf{J}(g \cdot (h \odot y)) \\ &= g \cdot \mathbf{J}(h \odot y) = g \cdot \mathbf{J}(y) = \mathbf{J}(g \cdot y)\ .\end{aligned}$$

That $\mathrm{T}\phi$ is symplectic on horizontal spaces now follows from C.1 and Lemma 8.12.

Conclusion 4 follows by construction. So does conclusion 5. Conclusion 6 was proven in C.2 above. It remains to prove that $\mathbf{K} : P \to \mathfrak{h}^*$ is a momentum map for the extended action of H.

Write η_P^H (resp. η_F^H) for the infinitesimal generator of the H action on P (resp. F) corresponding to $\eta \in \mathfrak{h}$. Let $\eta \in \mathfrak{h}$ be arbitrary. Then one computes using the definition of the extended action,

$$\text{C.3} \qquad \eta_P^H(g \cdot y) = \mathrm{T}\Phi_g \cdot \eta_F^H(y) \qquad (g \in G,\, y \in F)\ ,$$

where $\Phi_g(x) \equiv g \cdot x$ ($x \in P$). Consider the one-form $\alpha \equiv \eta_P^H \lrcorner\, \omega - dK_\eta$. Then as $\eta \in \mathfrak{h}$ is arbitrary, proving that $\mathbf{K} : P \to \mathfrak{h}^*$ is a momentum map for the extended action amounts to proving that α vanishes. It suffices to check

C.4 $$\alpha|\mathrm{T}_{g\cdot y}F_{g\cdot\mu_0}=0$$

C.5 $$\text{and}\qquad \alpha|\mathfrak{g}_P(g\cdot y)=0\ ,$$

where $g\in G$ and $y\in F$ are arbitrary, and $\mathfrak{g}_P(x)\equiv\mathrm{T}_x(G\cdot x)$ $(x\in P)$.

Suppose $v\in\mathrm{T}_{g\cdot y}F_{g\cdot\mu_0}$. Then $v=\mathrm{T}\Phi_g\cdot w$ for some $w\in\mathrm{T}_yF$. In that case we compute

$$\begin{aligned}\langle\alpha,v\rangle &= \omega\left(\eta_P^H(g\cdot y),\,\mathrm{T}\Phi_g\cdot w\right)-\langle dK_\eta,\mathrm{T}\Phi_g\cdot w\rangle\\ &= \omega\left(\mathrm{T}\Phi_g\cdot\eta_F^H(y),\,\mathrm{T}\Phi_g\cdot w\right)-\langle dK_\eta,w\rangle \quad \text{using C.3 and } \mathbf{K}\text{'s } G\text{-invariance}\\ &= \omega(\eta_F^H(y),w)-\langle dK_\eta,w\rangle \qquad \text{since } G \text{ acts symplectically}\\ &= 0\ ,\end{aligned}$$

since $\mathbf{K}:F\to\mathfrak{h}^*$ is a momentum map for the action of H on F, by hypothesis. This proves C.4.

Let $\xi\in\mathfrak{g}$ be arbitrary. Then, using C.3, the G-invariance of $\mathbf{K}$, and the symplecticity of Φ_g, one computes

$$\begin{aligned}\langle\alpha,\xi_P(g\cdot y)\rangle &= \langle\alpha,\mathrm{T}\Phi_g\cdot(g^{-1}\cdot\xi)_P(y)\rangle\\ &= \omega\left(\eta_F^H(y),\,(g^{-1}\cdot\xi)_P(y)\right)-\langle dK_\eta,(g^{-1}\cdot\xi)_P(y)\rangle\ .\end{aligned} \qquad \text{C.6}$$

By Lemma 8.12.1, we can write $g^{-1}\cdot\xi=\xi^1+\xi^2$ for some $\xi^1,\xi^2\in\mathfrak{g}$ such that $\xi_P^1(y)\in\mathrm{hor}_y$ and $\xi_P^2(y)\in\mathrm{T}_yF$. Then the right-hand side of C.6 becomes

$$\omega(\eta_F^H(y),\xi_P^1(y))-\langle dK_\eta,\xi_P^1(y)\rangle+\left(\omega(\eta_F^H(y),\xi_P^2(y))-\langle dK_\eta,\xi_P^2(y)\rangle\right)\ .$$

The first term vanishes, by the definition of hor_x. The second term vanishes because $\mathbf{K}$ is G-invariant. The term in parentheses vanishes because $\mathbf{K}:F\to\mathfrak{h}^*$ is the momentum map of the action of H on F, and because $\xi_P^2(y)\in\mathrm{T}F$. Therefore, $\langle\alpha,\xi_P(g\cdot y)\rangle=0$. Since $\xi\in\mathfrak{g}$ was arbitrary, equation C.5 must hold.

APPENDIX D

An Application of Converting Dynamic Integrability into Geometric Integrability: The Euler-Poinsot Rigid Body Revisited

The purpose of this appendix is to give an application of the results of Chap. 12. We show how the body symmetry of the axisymmetric Euler-Poinsot rigid body can be constructed *ab initio* by applying Theorem 12.6. The theorem can also be applied to obtain $\mathrm{SO}(3)\times S^1$ symmetry in the *asymmetric* rigid body, provided one restricts to appropriate open subsets of the phase space but we do not attempt this here.

Recall that, in the notation of Chap. 11, the Euler-Poinsot rigid body is described by the Hamiltonian system $(\mathrm{T}\,\mathrm{SO}(3), -d\Theta, H)$, where

D.1
$$H(\boldsymbol{\lambda}(\Lambda, m)) \equiv \frac{1}{2I_1}m_1^2 + \frac{1}{2I_2}m_2^2 + \frac{1}{2I_3}m_3^2$$

and Θ is the one-form on $\mathrm{T}\,\mathrm{SO}(3)$ defined by 11.2. Irrespective of the values of the moments of inertia I_1, I_2, I_3, the Hamiltonian H is invariant with respect to the action of $G \equiv \mathrm{SO}(3)$ defined by

D.2
$$A \cdot \boldsymbol{\lambda}(\Lambda, m) \equiv \boldsymbol{\lambda}(A\Lambda, m) \ ,$$

or equivalently by $A \cdot \boldsymbol{\rho}(\Lambda, n) \equiv \boldsymbol{\rho}(A\Lambda, An)$. This is just the action of $\mathrm{SO}(3)\times S^1$ of Chap. 11 with the S^1 part left out. A momentum map $\mathbf{J} : \mathrm{T}\,\mathrm{SO}(3) \to \mathbb{R}^3$ is given by

D.3
$$\mathbf{J}(\boldsymbol{\lambda}(\Lambda, m)) \equiv \Lambda m \ ,$$

or equivalently by $\mathbf{J}(\boldsymbol{\rho}(\Lambda, n)) \equiv n$.

Let $\mathrm{T}\,\mathrm{SO}(3)\backslash\mathbf{0}$ denote $\mathrm{T}\,\mathrm{SO}(3)$ with its zero section removed. Then G acts freely on $\mathrm{T}\,\mathrm{SO}(3)\backslash\mathbf{0}$ and $\mathbf{J}(\mathrm{T}\,\mathrm{SO}(3)\backslash\mathbf{0}) \subset \mathfrak{g}^*_{\mathrm{reg}} \cong \mathbb{R}^3\backslash\{0\}$. The Hamiltonian G-space

$$(\mathrm{T}\,\mathrm{SO}(3)\backslash\mathbf{0}, -d\Theta, \mathrm{SO}(3), \mathbf{J})$$

is dynamically integrable in the sense of 3.7 (p. 18).

Before turning to the axisymmetric case, let us study the symplectic cross-section of $\mathrm{T}\,\mathrm{SO}(3)\backslash\mathbf{0}$ in some detail.

The symplectic cross-section

Choose T, $\mathcal{W}$ and $\mathcal{O}$ as in Example 8.6. Then $\mathcal{W}$ and $\mathcal{O}$ intersect at the point $\mu_0 \equiv e_3$. The symplectic fibration $\pi : \mathrm{T}\,\mathrm{SO}(3)\backslash\mathbf{0} \to \mathcal{O} \cong S^2$ (see Theorem 8.8) is given by $\pi(\boldsymbol{\rho}(\Lambda, n)) = n/\left\|n\right\|$. The map $\psi : \mathrm{SO}(3) \times (0, \infty) \to \mathrm{T}\,\mathrm{SO}(3)\backslash\mathbf{0}$ defined by

D.4
$$\psi(\Lambda, p) \equiv \boldsymbol{\rho}(\Lambda^{-1}, pe_3)$$

is a diffeomorphism onto the symplectic cross-section $F \equiv \pi^{-1}(e_3)$ of the space

$$(\mathrm{T\,SO}(3)\backslash\mathbf{0}, -d\Theta, \mathrm{SO}(3), \mathbf{J})$$

(see Definition 8.9). The map ψ is symplectic if we equip $\mathrm{SO}(3) \times (0,\infty)$ with the symplectic structure $-d\Theta'$, where Θ' is the one-form on $\mathrm{SO}(3) \times (0,\infty)$ defined by the formula 11.3. The action of $T \cong S^1$ on $\mathrm{SO}(3) \times (0,\infty)$ that makes ψ : $\mathrm{SO}(3) \times (0,\infty) \xrightarrow{\sim} F$ an S^1-equivariant map is given by

D.5 $$\theta \cdot (\Lambda, p) \equiv (\Lambda e^{-\theta \hat{e}_3}, p) \ .$$

The action of $T \cong S^1$ on F has momentum map $\mathbf{J}^F : F \to \mathfrak{t}^* \cong \mathbb{R}$ given by $\mathbf{J}^F(\boldsymbol{\rho}(\Lambda, n)) = n \cdot e_3$. So $(\mathbf{J}^F \circ \psi)(\Lambda, p) = p$.

To summarize, $\psi : \mathrm{SO}(3) \times (0,\infty) \xrightarrow{\sim} F$ is an equivalence between the space

$$\big(\mathrm{SO}(3) \times (0,\infty), -d\Theta', S^1, \mathbf{J}^F \circ \psi\big)$$

and the symplectic cross-section F, where Θ', $\mathbf{J}^F \circ \psi$, and the action of S^1 are as described above.

Poisson reduction in the symplectic cross-section

Our next task is to obtain a concrete realization of the Poisson reduced space $F/T \cong (\mathrm{SO}(3) \times (0,\infty))/S^1$. First, observe that the symplectic structure $-d\Theta'$ on $\mathrm{SO}(3) \times (0,\infty)$ is $-\omega^*_{\mathrm{SO}(3)}$, where $\omega^*_{\mathrm{SO}(3)}$ is the symplectic structure that $\mathrm{SO}(3) \times (0,\infty)$ carries by virtue of being (up to the obvious identifications) the action-group model space $G \times \mathfrak{t}_0^*$ for $G = \mathrm{SO}(3)$ (see Lemma 9.4)[1].

In Example 10.8 we constructed action-group coordinates for geodesic motions on S^2, i.e., we constructed an equivalence of Hamiltonian SO(3)-spaces $\phi : \mathrm{SO}(3) \times (0,\infty) \xrightarrow{\sim} \mathrm{T}S^2\backslash\mathbf{0}$. This map, which was defined by

$$\phi(\Lambda, p) \equiv (\Lambda e_1, p\Lambda e_2) \ ,$$

is symplectic, when $\mathrm{SO}(3) \times (0,\infty)$ is equipped with the symplectic structure $\omega^*_{\mathrm{SO}(3)}$. But here we are equipping $\mathrm{SO}(3) \times (0,\infty)$ with the symplectic structure $-\omega^*_{\mathrm{SO}(3)}$. To make ϕ *symplectic* we therefore redefine the symplectic structure on $\mathrm{T}S^2\backslash\mathbf{0}$ to be $+d\Theta$, where Θ is the one-form on $\mathrm{T}S^2\backslash\mathbf{0}$ defined in 8.13.

We leave it to the reader to verify that the action of S^1 on $\mathrm{T}S^2\backslash\mathbf{0}$ with respect to which ϕ is S^1-equivariant is given by

D.6 $$\theta \cdot (q, v) \equiv (e^{-\theta\hat{\mu}} q, e^{-\theta\hat{\mu}} v) \ , \qquad \text{where} \qquad \mu \equiv q \times v / \|v\| \ .$$

The key observation at this point is that the orbits of this S^1 action are the fibers of the map $\mathbf{J} : \mathrm{T}S^2\backslash\mathbf{0} \to \mathbb{R}^3\backslash\{0\}$ defined by $\mathbf{J}(q, v) \equiv q \times v$ (not to be confused with the momentum map above of the same name). But we have already seen in Chap. 8 that $\mathbf{J}$ is the momentum map for a Hamiltonian action of SO(3) on $\mathrm{T}S^2\backslash\mathbf{0}$. Actually, since we have redefined the sign of the symplectic structure on $\mathrm{T}S^2\backslash\mathbf{0}$, the momentum map of this action is, in the current context, $-\mathbf{J}$. In particular, since $-\mathbf{J}$ is equivariant, it is *Poisson* if we equip $\mathbb{R}^3 \cong \mathfrak{so}(3)^*$ with the positive Lie-Poisson bracket $\{\,\cdot\,,\,\cdot\,\}_+$ (see, e.g., Marsden and Ratiu (1994, Chap. 10)). It follows that $\mathbf{J}$ is

[1]This is no mystery: The symplectic structure on $G \times \mathfrak{t}_0^*$ was *defined* by realizing $G \times \mathfrak{t}_0^*$ as a symplectic cross-section of T^*G (Chap. 9). The reason for the difference in sign observed here is the following: In the rigid body phase space $\mathrm{T}^*\,\mathrm{SO}(3)$ (which we have been identifying with $\mathrm{T\,SO}(3)$) the symmetry group SO(3) of H acts by cotangent lifting the left action on SO(3) defined by $g \cdot h \equiv gh$, while in Chap. 9 we cotangent lifted the left action on G defined by $g \cdot h \equiv hg^{-1}$.

Poisson if we equip $\mathbb{R}^3$ with the negative Lie-Poisson bracket $\{\,\cdot\,,\,\cdot\,\}_- \equiv -\{\,\cdot\,,\,\cdot\,\}_+$. This structure is given by

D.7 $$\{f,h\}_-(y) \equiv -y \cdot \nabla f(y) \times \nabla h(y) \ .$$

Since $\phi : \mathrm{SO}(3) \times (0,\infty) \to \mathrm{T}S^2\backslash\mathbf{0}$ is symplectic, the composite

$$\rho \equiv \mathbf{J} \circ \phi : \mathrm{SO}(3) \times (0,\infty) \to \mathbb{R}^3\backslash\{0\}$$

is also a Poisson map. This map is a surjective submersion whose fibers are the S^1-orbits in $\mathrm{SO}(3) \times (0,\infty)$ (by the S^1-equivariance of ϕ). Thus $(\mathbb{R}^3\backslash\{0\}, \{\,\cdot\,,\,\cdot\,\}_-)$ is a realization of the Poisson manifold $(\mathrm{SO}(3) \times (0,\infty))/S^1$, and $\rho : \mathrm{SO}(3) \times (0,\infty) \to \mathbb{R}^3\backslash\{0\}$ is a realization of the natural projection $\mathrm{SO}(3) \times (0,\infty) \to (\mathrm{SO}(3) \times (0,\infty))/S^1$.

Since $H|F$ is T-invariant ($T \cong S^1$), the Hamiltonian ψ^*H (which represents the restriction of H to the symplectic cross-section) is S^1-invariant, dropping via ρ to a function on $\mathbb{R}^3\backslash\{0\}$. Indeed, denoting the standard coordinate functions on $\mathbb{R}^3$ by y_1, y_2, y_3, we have $\psi^*H = h \circ \rho$, where $h : \mathbb{R}^3\backslash\{0\} \to \mathbb{R}$ is given by

$$h = \frac{1}{2I_1} y_1^2 + \frac{1}{2I_2} y_2^2 + \frac{1}{2I_3} y_3^2 \ .$$

The reader will readily verify this fact after observing that

$$\rho(\Lambda, p) = \mathbf{J}(\Lambda e_1, p\Lambda e_2) = \Lambda e_1 \times (p\Lambda e_2) = p\Lambda e_3 \ .$$

Hamiltonian vector fields on $(\mathbb{R}^3\backslash\{0\}, \{\,\cdot\,,\,\cdot\,\}_-)$ take the form

$$X_f(y) = -y \times \nabla f(y) \ ,$$

so that the equations of motion on $\mathbb{R}^3\backslash\{0\}$ corresponding to Hamiltonian h are

D.8 $$\begin{aligned} \dot{y}_1 &= \left(\frac{1}{I_2} - \frac{1}{I_3}\right) y_2 y_3 \\ \dot{y}_2 &= \left(\frac{1}{I_3} - \frac{1}{I_1}\right) y_3 y_1 \\ \dot{y}_3 &= \left(\frac{1}{I_1} - \frac{1}{I_2}\right) y_1 y_2 \ . \end{aligned}$$

These are precisely the familiar Lie-Poisson reduced rigid body equations (see, e.g., op. cit.), albeit obtained via a rather convoluted route: Rather than Poisson reduce the full system directly, we have obtained these equations by doing Poisson reduction in (a realization of) the symplectic cross-section.

The axisymmetric case

Before summarizing the above results, we need to restrict attention to an appropriate open subset of $\mathrm{T}\,\mathrm{SO}(3)\backslash\mathbf{0}$, in anticipation of applying Theorem 12.6.

Assume the body is axisymmetric, $I_1 = I_2 \equiv I$ say. Then, writing $z(t) \equiv y_1(t) + iy_2(t) \in \mathbb{C}$, the general solution to the equations D.8 is

D.9 $$y_3(t) = y_3(0)$$

D.10 $$z(t) = e^{i\sigma y_3(0)t} z(t) \ ,$$

D.11 $$\text{where} \quad \sigma \equiv \left(\frac{1}{I_3} - \frac{1}{I}\right) \ .$$

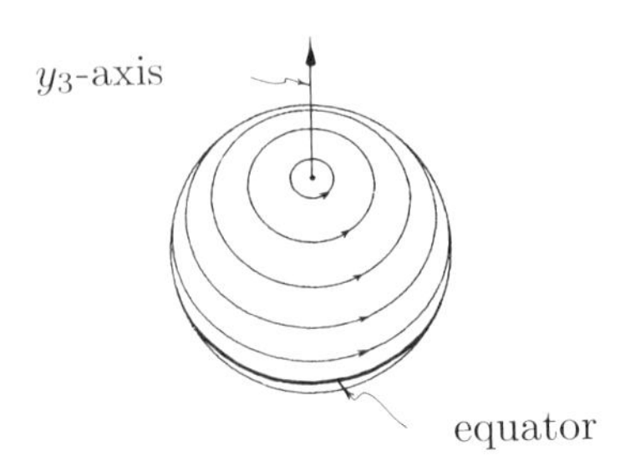

FIGURE 1. The reduced dynamics of an axisymmetric rigid body on a typical symplectic leaf (sphere). Points on the equator are equilibria.

Points on the y_3-axis or on the y_1-y_2 coordinate plane are equilibria. The remaining solutions are periodic with period $(\sigma y_3(0))^{-1}$. The symplectic leaves of $(\mathbb{R}^3\backslash\{0\}, \{\cdot, \cdot\}_-)$ are the spheres centered on the origin. The solution trajectories on a typical sphere are shown in Fig. 1. To apply Theorem 12.6 we need to restrict attention to the half-space lying above or below the y_1-y_2 coordinate plane. For concreteness, we choose the upper half-space $\mathbb{R}^3_+ \equiv \{y \in \mathbb{R}^3 \mid y_3 > 0\}$. Let us now summarize our previous results, with the various constructions restricted to the appropriate open subsets:

D.12 Proposition (Poisson reduction in the symplectic cross-section).
Define the open set $P \subset \mathrm{T}\,\mathrm{SO}(3)\backslash\mathbf{0}$ by

$$P \equiv \{\boldsymbol{\lambda}(\Lambda, m) \mid m_3 > 0\}$$

and equip P with the symplectic structure $\omega \equiv -d\Theta$, where Θ is the one-form on $\mathrm{T}\,\mathrm{SO}(3)$ defined by 11.2. Then P is invariant with respect to the action of $G = \mathrm{SO}(3)$ defined by D.2, which has momentum map $\mathbf{J} : P \to \mathbb{R}^3$ defined by D.3.

Define the open set $\mathrm{SO}(3)^\circ \subset \mathrm{SO}(3)$ by

$$\mathrm{SO}(3)^\circ \equiv \{\Lambda \in \mathrm{SO}(3) \mid \Lambda_{33} > 0\} \ ,$$

where $\Lambda_{33} \equiv \Lambda e_3 \cdot e_3$. Equip $F' \equiv \mathrm{SO}(3)^\circ \times (0, \infty)$ with the symplectic structure $\omega' \equiv -\omega^*_{\mathrm{SO}(3)}$, where $\omega^*_{\mathrm{SO}(3)}$ is the symplectic structure carried by $\mathrm{SO}(3) \times (0, \infty) \supset \mathrm{SO}(3)^\circ \times (0, \infty)$ by virtue of being (up to the obvious identifications) the action-group model space of $\mathrm{SO}(3)$ (see Lemma 9.4). Explicitly, $\omega' = d\Theta'$, where Θ' is the one-form on $\mathrm{SO}(3) \times (0, \infty)$ given in Equation 11.3. Let $T = S^1$ act on F' according to D.5. This action has momentum map $\mathbf{J}' : F' \to \mathbb{R}$ defined by $\mathbf{J}'(\Lambda, p) \equiv p$. Then the map $\psi : F' \to P$ defined by D.4 is an equivalence from $(F', \omega', T, \mathbf{J}')$ onto the symplectic cross-section $(F, \omega_F, T, \mathbf{J}^F)$ of $(P, \omega, G, \mathbf{J})$. The map $\rho : F' \to \mathbb{R}^3_+$ defined by

$$\rho(\Lambda, p) \equiv p\Lambda e_3$$

is a Poisson map onto $(\mathbb{R}^3_+, \{\cdot, \cdot\}_-)$, where $\{\cdot, \cdot\}_-$ is defined by D.7. Moreover, the fibers of ρ are the T-orbits in F', so that $(\mathbb{R}^3_+, \{\cdot, \cdot\}_-)$ is a realization of the Poisson manifold F'/T and ρ is a realization of the natural projection $F' \to F'/T$.

Assuming $I_1 = I_2 \equiv I$, *we have* $H' \equiv \psi^* H = h \circ \rho$, *where* $h : \mathbb{R}^3_+ \to \mathbb{R}$ *is defined by*

$$h \equiv \frac{1}{2I}(y_1^2 + y_2^2) + \frac{1}{2I_3} y_3^2 \ .$$

Applying Theorem 12.6

Let $(F', \omega', T, \mathbf{J}')$ be the realization of the symplectic cross-section F of $P \subset \mathrm{T\,SO}(3)$ given in the proposition above. The momentum map $\mathbf{J}'$ has image $U = (0, \infty)$, and $\mathbf{J}' : F' \to (0, \infty)$ factors through $\rho : F' \to \mathbb{R}^3_+$, delivering a map $\mathbf{j}' : \mathbb{R}^3_+ \to (0, \infty)$ such that $\mathbf{J}' = \mathbf{j}' \circ \rho$. Indeed $\mathbf{j}'(y) = \|y\|$. Each symplectic leaf $\Sigma_\mu = (\mathbf{j}')^{-1}(\mu)$ $(\mu \in (0, \infty))$ is an open hemisphere of radius μ.

If we take

$$Z \equiv \{(0, 0, t) \mid t > 0\} \ ,$$

then the Hamiltonian T-space $(F', \omega', T, \mathbf{J}')$ satisfies the hypotheses of Theorem 12.6. In the notation of 12.6, we have $\frac{1}{2\pi}\mathfrak{t}^{\mathbb{Z}} \cong \mathbb{Z}$; the function $I : \mathbb{R}^3_+ \to \mathbb{R}$ defined in 12.6 is given by

$$\begin{aligned} I(y) &= n\,\|y\| - A(y)/(2\pi\,\|y\|) \\ &= y_3 + (n-1)\,\|y\| \ , \end{aligned}$$

where $n \in \mathbb{Z}$ is arbitrary and we have set $c_0 \equiv 0$. In the first line of the formula $A(y) = 2\pi\,\|y\|\,(\|y\| - y_3)$ refers to the absolute area enclosed by the reduced orbit through y.

Since $\psi : F' \to F$ is an equivalence of Hamiltonian G-spaces, Theorem 12.6 implies that $I \circ \rho \circ \psi^{-1} : F \to \mathbb{R}$ is the momentum map for a new action of S^1 on the symplectic cross-section F commuting with the existing action of $T = S^1$ on F. If we take $H \equiv S^1$ and $\mathbf{K} \equiv I \circ \rho \circ \psi^{-1}$ in Lemma 8.17 (p. 51), then we may apply the lemma to extend our new S^1-action to a Hamiltonian action on P commuting with that of $G = \mathrm{SO}(3)$. In particular the lemma tells us how to extend $\mathbf{K}$ to a function on P that is a momentum map for the extended S^1-action. If one takes $n = 0$ then, as the reader is invited to verify, this extended momentum map is given by $\mathbf{K}(\lambda(\Lambda, m)) = m_3$. This is precisely the momentum map for the standard 'body' symmetry S^1-action, as written down and exploited in Chap. 10.

According to Remark 12.7 there are an infinite number of other ways to construct a new S^1-action on F, and hence P. One such action is generated by the 'new' action integral for the rigid body appearing in Tantalo (1994). This corresponds above to the choice $n = 1$ $[\,I(y) = -A(y)/(2\pi\,\|y\|)\,]$.

APPENDIX E

Dual Pairs, Leaf Correspondence, and Symplectic Reduction

The main result to be recalled here is a natural one-to-one correspondence between the set of symplectic leaves in each leg of a 'dual pair' (defined below). This result has been stated by Weinstein (1983), but to our knowledge no complete proof has ever appeared, although incorrect proofs exist in the literature. In view of this, we submit here a deliberately leisurely proof which includes frequent excursions into background material, counterexamples, etc.

The symplectic leaves of a Poisson manifold are the leaves of a foliation integrating a smooth generalized distribution on the manifold (the characteristic distribution). Our proof of the symplectic leaf correspondence theorem starts by understanding how generalized foliations (foliations with singularities) behave when 'pulled back' by a surjective submersion.

As an application of the correspondence theorem (as we shall formulate it) we establish the well-known equivalence of the three basic types of symplectic reduction: Marsden-Weinstein 'point' reduction, Marle 'orbit' reduction, and passage to a symplectic leaf of the Poisson-reduced space. Indeed, if μ is in the image of the momentum map, then the correspondence theorem establishes realizations of the corresponding point-reduced space P_μ and orbit-reduced space $P_{\mathcal{O}_\mu}$ ($\mathcal{O}_\mu$ the coadjoint orbit through μ) as the same symplectic leaf of the Poisson-reduced space (or 'pseudoleaf,' if the momentum map has disconnected fibers).

Outline. The symplectic leaf correspondence theorem proven here appears as Theorem E.13, together with Addendum E.14. Our applications to reduction are embodied in Theorems E.18 and E.17.

Throughout our discussion 'smooth' means C^∞. By a *manifold* we will always mean a smooth finite-dimensional manifold whose connected components share a common dimension.

Acknowledgments. We are grateful for helpful conversations with Tudor Ratiu, Juan-Pablo Ortega and Sergey Pekarsky.

Foliations with singularities

A *generalized distribution* on a smooth manifold P is a subset D of the tangent bundle $\mathrm{T}P$ with the property that $D(x) \equiv D \cap \mathrm{T}_xP$ is a subspace of T_xP at every $x \in P$. The adjective 'generalized' (which we will omit in the sequel) refers to the fact that we allow the dimension of $D(x)$ to be x-dependent. If X is a local vector field on P then, in a slight abuse of notation, we write $X \subset D$ if $X(x) \in D(x)$ at all $x \in P$ for which $X(x)$ is defined. A distribution D is deemed *smooth* if for all $x \in P$ there exist smooth vector fields $X_1, \dots, X_k \subset D$ such that $\{X_1(x), \dots, X_k(x)\}$ generates $D(x)$.

We are ultimately interested in the following special case:

E.1 Example. Let Q be a Poisson manifold and B the associated Poisson tensor, viewed as a vector bundle morphism $B : \mathrm{T}^*Q \to \mathrm{T}Q$. Next, define the *characteristic distribution* D on Q by $D \equiv B(\mathrm{T}^*Q)$. If $x_1, \dots, x_n$ are local coordinate functions defined in a neighborhood of a point $q \in Q$, then $D(q)$ is generated by $\{X_{x_1}(q), \dots, X_{x_n}(q)\}$, where $X_{x_j} \subset D$ is the smooth vector field defined by $X_{x_j}(x) \equiv B(d_x x_j)$. In particular, the characteristic distribution D is smooth.

Integral manifolds. Let D be a smooth distribution on a manifold P. An *integral manifold* Σ of D is the image of an injective immersion $i : \Sigma_0 \to P$ with the property $(\mathrm{T}_x i)(\mathrm{T}_x \Sigma_0) = D(i(x))$ for all $x \in \Sigma_0$. The map $i : \Sigma_0 \to P$, which is called a *realization* of Σ, transfers from Σ_0 to $\Sigma = i(\Sigma_0)$ a smooth manifold structure, with respect to which the inclusion $\Sigma \hookrightarrow P$ is an injective immersion. The integral manifold Σ is then equipped with a tangent space at each $x \in \Sigma$ which, viewed as a subspace of $\mathrm{T}_x P$, is $\mathrm{T}_x \Sigma = D(x)$.

E.2 Lemma. *If Σ is an integral manifold of a smooth distribution then any two realizations of Σ determine identical smooth manifold structures (and hence topologies) on Σ.*

PROOF. See the proof of Libermann and Marle (1987, Lemma 2.2, Appendix 3). □

We may thus speak of 'the' smooth manifold structure of an integral manifold. Similarly, by 'the' topology of an integral manifold we will always mean its topology in the sense of E.2 (which is not to be confused with the topology inherited from the ambient manifold). In particular, we call an integral manifold Σ *connected* if Σ_0 is connected in some, and hence all, realizations $i : \Sigma_0 \to P$ of Σ.

E.3 Warning. Despite Lemma E.2, integral manifolds can behave badly, especially if they are not connected. For example, a k-dimensional integral manifold could, as a subset of the ambient manifold, be a regular (i.e., embedded) submanifold of dimension different from k. Indeed, a manifold P supporting a distribution D can itself be an integral manifold for D, even if $\dim D(x) < \dim P$, for all $x \in P$:

E.4 Example. Take $P \equiv \mathbb{R}^2$ and let D be the smooth distribution generated by $\partial/\partial x_1$ (x_1 the first coordinate). Consider the paracompact one-dimensional manifold $\Sigma_0 \equiv \coprod_{s \in \mathbb{R}} \mathbb{R}$, where $\coprod$ denotes disjoint union, and let $i : \Sigma_0 \to \mathbb{R}^2$ be the map that sends a point t in the 'sth' connected component of Σ_0 to (s, t). Then i realizes $\mathbb{R}^2$ as an integral manifold of D.

We remind the reader that the analogue of E.2 fails for general immersed submanifolds, even if one restricts to *connected* realizations of a fixed dimension. A counterexample is given by the two realizations of the 'figure eight,' $i : (0, 2\pi) \to \mathbb{R}^2$ and $i' : (\pi, 3\pi) \to \mathbb{R}^2$, defined by $i(t) \equiv (\sin(t), \sin(2t)) \equiv i'(t)$. The induced map $\phi : (0, 2\pi) \to (\pi, 3\pi)$ has a discontinuity at π, so that i and i' determine different topologies on their common image.

Integrability. A distribution D is *integrable* if through every point $x \in P$ there exists an integral manifold of D. The following result is due to Sussmann

(1973) and Stefan (1973). See also Libermann and Marle (1987, Appendix 3) and the references contained therein.

E.5 Theorem and Definition. *If D is a smooth integrable distribution on a manifold P then for arbitrary points $x, y \in P$ the following are equivalent:*

1. *There exists a connected integral manifold of D containing both x and y.*
2. *There exist a finite number of smooth vector fields $X_1, \dots X_n \subset D$ with corresponding integral curves $\gamma_j : [0, t_j] \to P$ $(j = 1, \dots, n)$ that join x and y in the sense that $x = \gamma_1(0), \gamma_1(t_1) = \gamma_2(0), \gamma_2(t_2) = \gamma_3(0), \dots, \gamma_{n-1}(t_{n-1}) = \gamma_n(0), \gamma_n(t_n) = y$.*

Moreover, if one writes $x \sim y$ if 1 or 2 holds, then $\sim$ is an equivalence relation on P, defining a partition of P called the (generalized) foliation associated with D. The equivalence classes of $\sim$ are called the leaves of the foliation, and one has:

3. *Each leaf $\mathcal{L}$ is a maximal connected integral manifold in the sense that it is a connected integral manifold containing all other connected integral manifolds that pass through each of its points.*
4. *For each connected integral manifold $\Sigma \subset \mathcal{L}$ the inclusion $i : \Sigma \hookrightarrow \mathcal{L}$ is a diffeomorphism onto its image. In particular $\Sigma \subset \mathcal{L}$ is open.*

Note that since leaves are integral manifolds they have canonically defined smooth manifold structures; see E.2 above.

Notation. If D is an integrable distribution we denote the set of leaves of the associated foliation by $\mathcal{F}_D$.

Pull-backs. If $\rho : P \to Q$ is a smooth map, and D is a distribution on Q, then we define a distribution $\rho^* D$ on P by declaring $v \in (\rho^* D)(x)$ if $\mathrm{T}\rho \cdot v \in D(\rho(x))$. We call $\rho^* D$ the *pull-back* of D. The next three results describe how distributions, and foliations associated with integrable distributions, behave when pulled back by a submersion. These results will be crucial in our proof of the correspondence theorem. Throughout D denotes a fixed distribution on a manifold Q.

E.6 Lemma. *If $\rho : P \to Q$ is a submersion and D is smooth then so is $\rho^* D$. If additionally D is integrable then so is $\rho^* D$.*

E.7 Theorem. *Suppose that $\rho : P \to Q$ is a surjective submersion and that D is smooth and integrable. Then for each leaf $\mathcal{L} \subset Q$ of the foliation associated with D the set $\rho^{-1}(\mathcal{L}) \subset P$ is a (possibly disconnected) integral manifold of $\rho^* D$ whose connected components are leaves of the foliation associated with $\rho^* D$. The restriction $\rho : \rho^{-1}(\mathcal{L}) \to \mathcal{L}$ is a submersion, with respect to the smooth manifold structures of the domain and range as integral manifolds (see E.2). If $\rho : P \to Q$ has connected fibers, then $\rho^{-1}(\mathcal{L})$ is a single leaf of the foliation associated with $\rho^* D$.*

Lemma E.6 and Theorem E.7 are proven at the end of this Appendix, under 'Details.' As foliations are partitions, Theorem E.7 implies the following:

E.8 Corollary. *Assume $\rho : P \to Q$ is a surjective submersion with connected fibers and assume that D is smooth and integrable. Then for each leaf $\mathcal{K} \in \mathcal{F}_{\rho^* D}$, the set $\rho(\mathcal{K})$ is a leaf in $\mathcal{F}_D$ and the map*

$$\mathcal{K} \mapsto \rho(\mathcal{K}) : \mathcal{F}_{\rho^* D} \to \mathcal{F}_D$$

is a bijection with inverse

$$\mathcal{L} \mapsto \rho^{-1}(\mathcal{L}) \ .$$

The following example demonstrates the necessity of fiber connectedness in the last part of Theorem E.7, and in the corollary. In fact, this example shows that even if the fibers of $\rho : P \to Q$ are finite in the number of their connected components, $\rho^{-1}(\mathcal{L})$ ($\mathcal{L} \in \mathcal{F}_D$) may be an *infinite* union of leaves $\mathcal{K} \in \mathcal{F}_{\rho^* D}$. It also shows that without connectedness the image $\rho(\mathcal{K})$ of a leaf $\mathcal{K} \in \mathcal{F}_{\rho^* D}$ need not be an entire leaf $\mathcal{L} \in \mathcal{F}_D$.

E.9 Example. Let P be the open strip $\{(x_1, x_2) \in \mathbb{R}^2 \mid 0 < x_2 < 1\}$ with sloping line segments removed as indicated in Fig. 1. Take $Q \equiv \mathbb{R}$ and define $\rho : P \to Q$ to be the surjective submersion defined by $\rho(x_1, x_2) \equiv x_1$. Take $D \equiv \mathrm{T}Q$ so that $\mathcal{F}_D$ consists of a single leaf, namely all of Q. Then $\rho^* D = \mathrm{T}P$ so that the leaves in $\mathcal{F}_{\rho^* D}$ are the connected components of P (by E.7), i.e., they are the parallelogram regions $\dots, \mathcal{L}_{-1}, \mathcal{L}_0, \mathcal{L}_1, \mathcal{L}_2, \dots$ indicated in the figure.

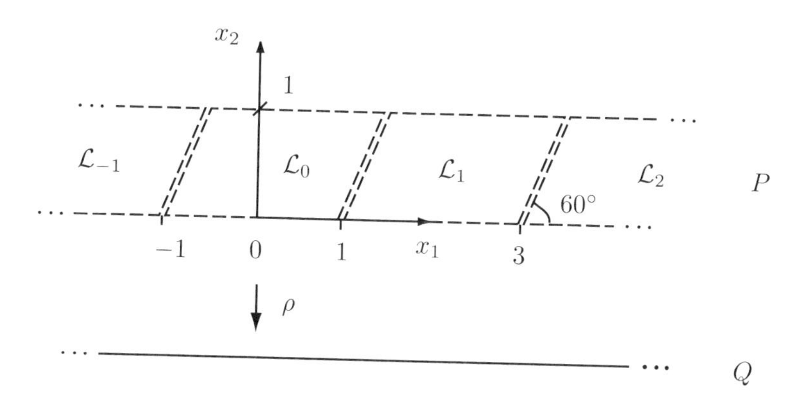

FIGURE 1. A counterexample (see Example E.9) for the connectedness hypothesis in Theorem E.7 and Corollary E.8.

The correspondence theorem

Symplectic leaves. We now recall Kirillov's (1976) well known result expressing a Poisson manifold as a disjoint union of symplectic manifolds called the *symplectic leaves* (see also Libermann and Marle (1987, p. 130)):

E.10 Theorem (Symplectic stratification theorem).
The characteristic distribution on a Poisson manifold Q (see Example E.1 above) is integrable. Furthermore, each leaf $\mathcal{L}$ of the associated foliation inherits a symplectic structure, with respect to which the inclusion $\mathcal{L} \hookrightarrow P$ is a Poisson map.

E.11 Lemma. *Let P be a symplectic manifold with symplectic structure ω and $\rho : P \to Q$ a Poisson submersion onto a Poisson manifold Q. If D denotes the characteristic distribution on Q, then*

$$\rho^* D = \ker \mathrm{T}\rho + (\ker \mathrm{T}\rho)^{\omega} \ .$$

Also, if $x \in P$ is arbitrary, $\mathcal{L} \subset Q$ is the symplectic leaf passing through $\rho(x)$, and $\omega_{\mathcal{L}}$ denotes the symplectic structure of $\mathcal{L}$, then

$$\omega(u, v) = \omega_{\mathcal{L}}(\mathrm{T}\rho \cdot u, \mathrm{T}\rho \cdot v) \quad \text{for all} \quad u, v \in (\ker \mathrm{T}_x\rho)^{\omega} \ .$$

PROOF. The proof of the first formula is a routine exercise in applying the definition of a Poisson map, as characterized by the Poisson tensor. Indeed, let $B_P : \mathrm{T}^*P \to \mathrm{T}P$ and $B_Q : \mathrm{T}^*Q \to \mathrm{T}Q$ denote the Poisson tensors. Let $\rho_* : \operatorname{Ann} \ker \mathrm{T}\rho \to \mathrm{T}^*Q$ denote the map sending $d_x(f \circ \rho)$ to $d_{\rho(x)}f$, for any locally defined function f on Q (Ann denotes annihilator). This map is well defined because ρ is a submersion. Then $\rho : P \to Q$ being Poisson means $B_Q \circ \rho_* = \mathrm{T}\rho \circ B_P$. Fix $x \in P$. Then

$$\begin{aligned} \mathrm{T}_x\rho\,(\,(\ker \mathrm{T}_x\rho)^{\omega}\,) &= \mathrm{T}_x\rho\,(\,B_P(\operatorname{Ann}\ker \mathrm{T}_x\rho)\,) \\ &= B_Q\,(\,\rho_*(\operatorname{Ann}\ker \mathrm{T}_x\rho)\,) = B_Q(\mathrm{T}^*_{\rho(x)}Q) = D(\rho(x)) \ . \end{aligned}$$

Here we have used the fact that ρ_* maps $\operatorname{Ann}\ker \mathrm{T}_x\rho$ *onto* (in fact isomorphically onto) $\mathrm{T}^*_{\rho(x)}Q$, which follows from the definition of ρ_*.

The above calculation demonstrates that $(\mathrm{T}_x\rho)^{-1}(D(\rho(x))) = (\ker \mathrm{T}_x\rho) + (\ker \mathrm{T}_x\rho)^{\omega}$ $(x \in P)$, i.e., that the first formula in the lemma holds.

Regarding the second formula, observe that arbitrary vectors $u, v \in (\ker \mathrm{T}_x\rho)^{\omega}$ are of the form $u = X_{f\circ\rho}(x), v = X_{h\circ\rho}(x)$ for some locally defined functions f, h on Q. This is because $(\ker \mathrm{T}_x\rho)^{\omega} = B_P(\operatorname{Ann}\ker \mathrm{T}_x\rho)$. Consequently,

$$\begin{aligned} \omega(u, v) &= \omega(X_{f\circ\rho}(x), X_{h\circ\rho}(x)) = \{f \circ \rho, h \circ \rho\}(x) \\ &= \{f, h\}(\rho(x)) = \omega_{\mathcal{L}}(X_f(\rho(x)), X_h(\rho(x))) = \omega_{\mathcal{L}}(\mathrm{T}\rho \cdot u, \mathrm{T}\rho \cdot v) \ . \end{aligned}$$

□

Dual pairs. Let P be a symplectic manifold and Q_1, Q_2 Poisson manifolds. A pair of Poisson maps $Q_1 \xleftarrow{\rho_1} P \xrightarrow{\rho_2} Q_2$ is called a *dual pair* if $\ker \mathrm{T}\rho_1$ and $\ker \mathrm{T}\rho_2$ are symplectically orthogonal distributions. This pair is a *full* dual pair if ρ_1 and ρ_2 are surjective submersions.

E.12 Example (The setting of symplectic reduction). Suppose a Lie group G acts on P in a Hamiltonian fashion, and let $\mathbf{J} : P \to \mathfrak{g}^*$ be a momentum map. For simplicity, assume that $\mathbf{J}$ is Ad^*-equivariant[1]. Then, assuming that G acts freely and properly, the image $\mathbf{J}(P)$ of $\mathbf{J}$ is an open subset of $\mathfrak{g}^*$, and $\mathbf{J}(P) \xleftarrow{\mathbf{J}} P \to P/G$ is a full dual pair, with respect to an appropriate differentiable structure on P/G.

We are now ready to state the symplectic leaf correspondence theorem for dual pairs. It is an immediate corollary of the first part of Lemma E.11, and Corollary E.8.

E.13 Theorem (The symplectic leaf correspondence theorem). *If $Q_1 \xleftarrow{\rho_1} P \xrightarrow{\rho_2} Q_2$ is a full dual pair, then the distribution $E \equiv \ker \mathrm{T}\rho_1 + \ker \mathrm{T}\rho_2$ is simultaneously the pullback of the characteristic distribution D_1 on Q_1 and of the characteristic distribution D_2 on Q_2. Moreover, if ρ_1 and ρ_2 have connected fibers, then one has bijections*

[1]Otherwise, replace the co-adjoint action by the action described in, e.g., Abraham and Marsden (1978, Proposition 4.2.7), and in the sequel replace 'co-adjoint orbit' by 'G-orbit.'

$$\begin{array}{ccccccc} \mathcal{F}_1 & \cong & & \mathcal{F}_E & & \cong & \mathcal{F}_2 \\ \rho_1(\mathcal{K}) & \leftarrow\!\shortmid & \mathcal{K} & & \mathcal{K} & \mapsto & \rho_2(\mathcal{K}) \\ \mathcal{L} & \mapsto & \rho_1^{-1}(\mathcal{L}) & & \rho_2^{-1}(\mathcal{L}) & \leftarrow\!\shortmid & \mathcal{L} \end{array}$$

where $\mathcal{F}_1 \equiv \mathcal{F}_{D_1}$ (resp. $\mathcal{F}_2 \equiv \mathcal{F}_{D_2}$) is the set of symplectic leaves in Q_1 (resp. Q_2) and $\mathcal{F}_E$ is the set of leaves of the foliation associated with E. In particular, the map $\mathcal{L} \mapsto \rho_2(\rho_1^{-1}(\mathcal{L})) : \mathcal{F}_1 \to \mathcal{F}_2$, which has inverse $\mathcal{L} \mapsto \rho_1(\rho_2^{-1}(\mathcal{L}))$, establishes a one-to-one correspondence between the symplectic leaves of Q_1 and those of Q_2. Corresponding leaves have equal codimension.

Suppose $\mathcal{L}_1 \in \mathcal{F}_1$, $\mathcal{K} \in \mathcal{F}_E$ and $\mathcal{L}_2 \in \mathcal{F}_2$ are leaves in correspondence, in the sense of the bijections of Theorem E.13, so that we have $\rho_1(\mathcal{K}) = \mathcal{L}_1$, $\rho_2(\mathcal{K}) = \mathcal{L}_2$ and $\rho_1^{-1}(\mathcal{L}_1) = \mathcal{K} = \rho_2^{-1}(\mathcal{L}_2)$. We claim the symplectic structures $\omega_{\mathcal{L}_1}$, $\omega_{\mathcal{L}_2}$ and ω of, respectively, $\mathcal{L}_1$, $\mathcal{L}_2$ and P, are related. Indeed, let $u, v \in E(x)$ be arbitrary ($x \in P$). Then $u = u_1 + u_2$ and $v = v_1 + v_2$ for some $u_1, v_1 \in \ker \mathrm{T}\rho_1$ and $u_2, v_2 \in \ker \mathrm{T}\rho_2$. Because $\ker \mathrm{T}\rho_1$ and $\ker \mathrm{T}\rho_2$ are ω-orthogonal, we compute, with the aid of the second formula in E.11,

$$\begin{aligned} \omega(u, v) &= \omega(u_1, v_1) + \omega(u_2, v_2) \\ &= \omega_{\mathcal{L}_2}(\mathrm{T}\rho_2 \cdot u_1, \mathrm{T}\rho_2 \cdot v_1) + \omega_{\mathcal{L}_1}(\mathrm{T}\rho_1 \cdot u_2, \mathrm{T}\rho_1 \cdot v_2) \\ &= \omega_{\mathcal{L}_2}(\mathrm{T}\rho_2 \cdot u, \mathrm{T}\rho_2 \cdot v) + \omega_{\mathcal{L}_1}(\mathrm{T}\rho_1 \cdot u, \mathrm{T}\rho_1 \cdot v) \ . \end{aligned}$$

Theorem E.7 and the preceding calculation establishes the following

E.14 Addendum (On the symplectic structures of corresponding leaves). *Let $\mathcal{L}_1 \in \mathcal{F}_1$, $\mathcal{K} \in \mathcal{F}_E$ and $\mathcal{L}_2 \in \mathcal{F}_2$ be leaves in correspondence, in the sense of Theorem E.13. Then the restrictions $\rho_1|_{\mathcal{K}} : \mathcal{K} \to \mathcal{L}_1$ and $\rho_2|_{\mathcal{K}} : \mathcal{K} \to \mathcal{L}_2$ are surjective submersions and*

$$i_{\mathcal{K}}^*\omega = \rho_1|_{\mathcal{K}}^*\omega_{\mathcal{L}_1} + \rho_2|_{\mathcal{K}}^*\omega_{\mathcal{L}_2} \ .$$

Here $i_{\mathcal{K}} : \mathcal{K} \hookrightarrow P$ denotes inclusion and ω, $\omega_{\mathcal{L}_1}$, $\omega_{\mathcal{L}_2}$ denote the symplectic structures on P, $\mathcal{L}_1$, $\mathcal{L}_2$ respectively.

Generalizations. More generally, we may consider a full dual pair in which only one leg has connected fibers:

E.15 Theorem (Correspondence in dual pairs with a disconnected leg). *Let $Q_1 \xleftarrow{\rho_1} P \xrightarrow{\rho_2} Q_2$ be a full dual pair and assume ρ_2 has connected fibers. Call sets of the form $\rho_2(\rho_1^{-1}(\mathcal{L}))$ for some $\mathcal{L} \in \mathcal{F}_1$ pseudoleaves of Q_2. Then:*

1. *Each pseudoleaf is a (possibly disconnected) integral manifold of the characteristic distribution D_2 of Q_2 and inherits a symplectic structure with respect to which the inclusion into Q_2 is a Poisson map. Moreover, the connected components of a pseudoleaf are symplectic leaves of Q_2.*
2. *The collection of all pseudoleaves constitutes a partition $\mathcal{F}_2'$ of Q_2. Moreover, Theorem E.13 and its addendum E.14 hold, provided $\mathcal{F}_2$ is replaced by $\mathcal{F}_2'$. In particular, there is a one-to-one correspondence between the symplectic leaves of Q_1 and the pseudoleaves of Q_2.*

E.16 Remark. It is important to emphasize that pseudoleaves need not be connected. In particular, Warning E.3 should be kept in mind. Also, in light of Example E.9, we should not rule out pseudoleaves with infinitely many connected components, even if the fibers of ρ_1 are finite in the number of their connected components. Indeed, without a second countability hypothesis, or something like it, on the symplectic leaves of Q_1 (and the manifold P), it is not clear how to exclude an *uncountably* infinite number of connected components,[2] as in the integral manifold of Example E.4.

PROOF OF THEOREM E.15. By applying Theorem E.7 to the first leg $\rho_1 : P \to Q_1$ of the pair, we know that each set $\rho_1^{-1}(\mathcal{L})$ ($\mathcal{L} \in \mathcal{F}_1$) is an integral manifold of E whose connected components are leaves of the associated foliation. It follows, applying Corollary E.8 to the other leg, whose fibers are connected, that $\rho_2(\rho_1^{-1}(\mathcal{L}))$ is then a union of symplectic leaves of Q_2. This proves part 1 of the theorem. Part 2 follows easily from E.7, E.8 and E.11. □

Symplectic reduction

If G in Example E.12 is connected, then $\mathbf{J}(P) \xleftarrow{\mathbf{J}} P \to P/G$ is a full dual pair whose second leg $P \to P/G$ has connected fibers. Applying Theorem E.15 to this case, and writing out the conclusions of E.13 and E.14 (with $\mathcal{F}_2$ replaced with the foliation $\mathcal{F}_2'$ by pseudoleaves) gives the following version of orbit reduction:

E.17 Theorem (Orbit-reduced spaces as pseudoleaves). *Assume that G is connected in Example E.12 and let $\mathcal{O} \subset \mathbf{J}(P)$ be an arbitrary co-adjoint orbit. Then:*

1. *$\mathbf{J}^{-1}(\mathcal{O})$ is an integral manifold of the generalized distribution E on P defined by $E(x) \equiv \ker \mathrm{T}_x\mathbf{J} + \mathrm{T}_x(G \cdot x)$.*
2. *The quotient space $P_\mathcal{O} \equiv \mathbf{J}^{-1}(\mathcal{O})/G$ is, as a subset of P/G, a pseudoleaf, in the sense of Theorem E.15. In particular $P_\mathcal{O}$ is a (possibly disconnected) integral manifold of the characteristic distribution on P/G and possesses a symplectic structure $\omega_{P_\mathcal{O}}$, with respect to which the inclusion $P_\mathcal{O} \hookrightarrow P/G$ is a Poisson map. The connected components of $P_\mathcal{O}$ are symplectic leaves; if $\mathbf{J} : P \to \mathbf{J}(P)$ has connected fibers, then $P_\mathcal{O}$ is a single symplectic leaf.*
3. *The restriction $\mathbf{J}_\mathcal{O} : \mathbf{J}^{-1}(\mathcal{O}) \to \mathcal{O}$ of $\mathbf{J}$ and the quotient map $\pi_\mathcal{O} : \mathbf{J}^{-1}(\mathcal{O}) \xrightarrow{/G} P_\mathcal{O}$ are surjective submersions, with respect to the smooth structure of domain and range as integral manifolds (see E.2).*
4. *If $\omega_\mathcal{O}^-$ denotes the 'minus' co-adjoint orbit symplectic structure on $\mathcal{O}$ and ω denotes the symplectic structure of P, then*
$$i_\mathcal{O}^*\omega = \mathbf{J}_\mathcal{O}^*\omega_\mathcal{O}^- + \pi_\mathcal{O}^*\omega_{P_\mathcal{O}} \ .$$
Here $i_\mathcal{O} : \mathbf{J}^{-1}(\mathcal{O}) \hookrightarrow P$ denotes inclusion.
5. *$P_\mathcal{O} = P_{\mathcal{O}'}$ if and only if $\mathcal{O} = \mathcal{O}'$. Also, we have $P/G = \cup_\mathcal{O} P_\mathcal{O}$, where the union is over all co-adjoint orbits $\mathcal{O} \subset \mathbf{J}(P)$, and is disjoint. One has* $\operatorname{codim} P_\mathcal{O} = \operatorname{codim} \mathcal{O}$.

That $P_\mathcal{O} = \mathbf{J}^{-1}(\mathcal{O})/G$ is a smooth manifold admitting a symplectic structure $\omega_{P_\mathcal{O}}$ satisfying the formula in E.17.4, is Marle's (1976) orbit reduction theorem.

Like orbit-reduced spaces, point-reduced spaces can also be realized as pseudoleaves of P/G:

---

[2] Note that second countability of Q_1 does not imply second countability of the symplectic leaves; these leaves are merely immersed submanifolds and need not possess the relative topology.

E.18 Theorem (Point-reduced spaces as pseudoleaves). *Assume that G in Example E.12 is connected. Let $\mu \in \mathbf{J}(P)$ be arbitrary, let $\mathcal{O}$ denote the co-adjoint orbit through μ, and let P_μ denote the image of $\mathbf{J}^{-1}(\mu)$ under the natural projection $P \to P/G$. Then:*

1. *The set P_μ coincides with the pseudoleaf $P_{\mathcal{O}} \subset P/G$ of Theorem E.17. In particular, P_μ is a (possibly disconnected) integral manifold of the characteristic distribution on P/G and possesses a symplectic structure ω_μ, with respect to which the inclusion $P_\mu \hookrightarrow P/G$ is a Poisson map. The connected components of P_μ are symplectic leaves; if $\mathbf{J} : P \to \mathbf{J}(P)$ has connected fibers, then P_μ is a single symplectic leaf.*
2. *The restriction $\pi_\mu : \mathbf{J}^{-1}(\mu) \to P_\mu$ of the natural projection $P \to P/G$ is a surjective submersion whose fibers are G_μ-orbits in P, i.e., P_μ is a realization of the abstract quotient $\mathbf{J}^{-1}(\mu)/G_\mu$.*
3. *If $i_\mu : \mathbf{J}^{-1}(\mu) \hookrightarrow P$ denotes the inclusion, then*
$$i_\mu^* \omega = \pi_\mu^* \omega_\mu \ .$$
4. *$P_\mu \cap P_{\mu'} \neq \emptyset$ if and only if $P_\mu = P_{\mu'}$, which is true if and only if μ and μ' lie on the same co-adjoint orbit. Also, $P/G = \cup_{\mu \in \mathfrak{g}^*_{\mathbf{J}}} P_\mu$ and* codim $P_\mu =$ codim $\mathcal{O}$.

That $\mathbf{J}^{-1}(\mu)/G_\mu$ is a smooth manifold admitting a symplectic structure ω_μ satisfying the equation in E.18.3 ($\pi_\mu : \mathbf{J}^{-1}(\mu) \to \mathbf{J}^{-1}(\mu)/G_\mu$ the abstract quotient map), is the point-reduction theorem of Marsden and Weinstein (1974) and Meyer (1973).

Proof. Let $\pi : P \to P/G$ denote the natural projection. It is not difficult to show that momentum map equivariance implies that $\pi(\mathbf{J}^{-1}(\mu)) = \pi(\mathbf{J}^{-1}(\mathcal{O}))$, i.e., that $P_\mu = P_{\mathcal{O}}$. This proves E.17.1.

Let us now establish that $\pi_\mu : \mathbf{J}^{-1}(\mu) \to P_\mu$ is a smooth map of manifolds (P_μ being equipped with the smooth manifold structure it possesses as an integral manifold). The map $\pi_\mu : \mathbf{J}^{-1}(\mu) \to P_\mu$ may be viewed as the restriction of the submersion $\pi_{\mathcal{O}} : \mathbf{J}^{-1}(\mathcal{O}) \to P_{\mathcal{O}} = P_\mu$ in E.17.3. So to prove π_μ is smooth, it suffices to establish that $\mathbf{J}^{-1}(\mu)$ is a regular submanifold of $\mathbf{J}^{-1}(\mathcal{O})$ (not just of P/G). But this follows from the preimage theorem, applied to the submersion $\mathbf{J}_{\mathcal{O}} : \mathbf{J}^{-1}(\mathcal{O}) \to \mathcal{O}$ of E.17.3.

The distribution E in E.17.1 is, by Theorem E.13, the pullback under π of the characteristic distribution on P/G, which has P_μ as an integral manifold. This implies that $(\mathrm{T}\pi)(E(x)) = \mathrm{T}_x P_\mu$, for any $x \in \mathbf{J}^{-1}(\mu)$. Since $(\mathrm{T}\pi_\mu)(\ker \mathrm{T}_x\mathbf{J}) = (\mathrm{T}\pi)(\ker \mathrm{T}_x\mathbf{J}) = (\mathrm{T}\pi)(E(x))$, this proves that π_μ is a submersion; π_μ is surjective by the definition of P_μ. That the fibers of π_μ are G_μ-orbits follows from momentum map equivariance. This completes the proof of E.18.2. Part 3 follows from E.11; part 4 follows from E.17.5. □

Details

Proof of E.6. Assume D is smooth. To show that $\rho^* D$ is smooth we may work locally which, on account of the submersion theorem, reduces the problem to the case where $P \subset \mathbb{R}^{p-q} \times \mathbb{R}^q$ and $Q \subset \mathbb{R}^q$ are open subsets ($0 \leqslant q \leqslant p$) and $\rho(\xi, \eta) = \eta$. Let $x = (\xi, \eta) \in P \subset \mathbb{R}^{p-q} \times \mathbb{R}^q$ be given. As D is smooth, there exist smooth locally defined vector fields $Y_1, \dots Y_k$ on $\mathbb{R}^q$ such that $\{Y_1(\eta), \dots, Y_k(\eta)\}$ generates $D(\eta)$. Let $\xi_1, \dots \xi_{p-q}, \eta_1, \dots, \eta_q$ denote the standard coordinate functions on $\mathbb{R}^{p-q} \times \mathbb{R}^q$. Without loss of generality, suppose the Y_j ($0 \leqslant j \leqslant k$)

are defined on all of Q, and lift the Y_j to smooth vector fields $\tilde{Y}_j$ (ρ-related to the Y_j) defined through $\langle d\xi_i, \tilde{Y}_j \rangle = 0, \langle d\eta_i, \tilde{Y}_j \rangle = \delta_{ij}$. Then the vector fields $\partial/\partial\xi_1, \ldots, \partial/\partial\xi_{p-q}, \tilde{Y}_1, \ldots \tilde{Y}_k$ lie in ρ^*D and, evaluated at x, generate $(\rho^*D)(x)$. Whence ρ^*D is smooth.

Assume D is smooth and integrable. Let $x_0 \in P$ be arbitrary and let $\mathcal{L} \in \mathcal{F}_D$ denote the leaf passing through $y_0 \equiv \rho(x_0)$. Next, let $U \subset \mathcal{L}$ be an open neighborhood of y_0 small enough to be a regular (i.e., embedded) submanifold of Q. Then by the transversal mapping theorem $\rho^{-1}(U)$ is a regular submanifold whose tangent space at a point x is $(\mathrm{T}_x\rho)^{-1}(\mathrm{T}_{\rho(x)}U) = (\rho^*D)(x)$. In other words, $\rho^{-1}(U)$ is an integral manifold of ρ^*D passing through x_0. So ρ^*D is integrable. □

In the proof of Theorem E.7 we will make use of the following result:

E.19 Proposition (A transversal mapping theorem for immersed submanifolds). *If $\rho : P \to Q$ is a smooth map transverse to an immersed submanifold Σ (i.e., $(\mathrm{T}_x\rho)(\mathrm{T}_xP) + \mathrm{T}_{\rho(x)}\Sigma = \mathrm{T}_{\rho(x)}Q$ for all $x \in \rho^{-1}(\Sigma)$), then there is a smooth manifold structure on $\rho^{-1}(\Sigma)$, with respect to which the inclusion $\rho^{-1}(\Sigma) \hookrightarrow P$ is an immersion, and such that the restriction $\rho : \rho^{-1}(\Sigma) \to \Sigma$ is a submersion. The tangent space of $\rho^{-1}(\Sigma)$ at a point x is $(\mathrm{T}_x\rho)^{-1}(\mathrm{T}_{\rho(x)}\Sigma)$.*

Proof. Cover Σ by open subsets U_i ($i \in I$) small enough that they are regular (i.e., embedded) submanifolds of P. By the transversal mapping theorem for regular submanifolds (see, e.g., Abraham et al. (1988, Theorem 3.5.12)), each set $M_i \equiv \rho^{-1}(U_i)$ is a regular submanifold of P whose tangent space at a point x is $(\mathrm{T}_x\rho)^{-1}(\mathrm{T}_{\rho(x)}\Sigma)$. The set $\rho^{-1}(\Sigma)$ is thus a union $\cup_{i\in I} M_i$ of regular submanifolds $M_i \subset P$ with the property that $x \in M_i \cap M_j \Rightarrow \mathrm{T}_xM_i = \mathrm{T}_xM_j$. But it is a fact that any two regular submanifolds M_i and M_j of a manifold which satisfy this last property induce identical smooth manifold structures on their intersection $M_i \cap M_j$, which is an open subset of both of them. It follows that if $\mathcal{A}_i$ is a representative atlas of coordinate charts for each M_i ($i \in I$), then $\cup_{i\in I} \mathcal{A}_i$ is a bona fide atlas of charts for $\rho^{-1}(\Sigma) = \cup_{i\in I} M_i$. This atlas gives $\rho^{-1}(\Sigma)$ the structure of a smooth manifold possessing the properties asserted in the proposition. □

Proof of Theorem E.7. Let $\mathcal{F}_D$ (resp. $\mathcal{F}_{\rho^*D}$) denote the set of leaves of the foliation associated with D (resp. ρ^*D). Let $\mathcal{L} \in \mathcal{F}_D$ be a leaf. Then as $\rho : P \to Q$ is a submersion, it is transverse to $\mathcal{L}$. It follows from Proposition E.19 that the set $\rho^{-1}(\mathcal{L})$ is an integral manifold of ρ^*D, and that the restriction $\rho : \rho^{-1}(\mathcal{L}) \to \mathcal{L}$ is a submersion, as claimed in the theorem statement. The remaining assertions of the theorem will follow once the claim below is established.

Claim. For any leaf $\mathcal{K} \in \mathcal{F}_{\rho^*D}$, the image $\rho(\mathcal{K})$ is an open subset of a leaf in $\mathcal{F}_D$.

Indeed, let $x \in \mathcal{K}$ be arbitrary and let $U \subset \mathcal{K}$ be a connected open subset that is small enough to be a regular (i.e., embedded) submanifold of P. As $\mathcal{K}$, and hence U, is an integral manifold of $\rho^*D \supset \ker \mathrm{T}\rho$, it follows that $\ker \mathrm{T}_x(\rho|U) = \ker \mathrm{T}_x\rho$ ($x \in U$). In particular, $\mathrm{T}_x(\rho|U)$ has constant rank on U. Applying the local subimmersion theorem (see, e.g., Abraham et al. (1988, Proposition 3.5.16)), we may conclude, shrinking U if necessary, that $\rho(U)$ is a connected submanifold of Q whose tangent spaces at a point $y = \rho(x)$ is $(\mathrm{T}_x\rho)(\mathrm{T}_xU) = (\mathrm{T}_x\rho)((\rho^*D)(x)) = D(y)$. In other words, $\rho(U)$ is a connected integral manifold of D. By parts 3 and 4 of Theorem E.5, $\rho(U)$ is an open subset of a leaf in $\mathcal{F}_D$. Since the point $x \in \mathcal{K}$

above is arbitrary, we may cover $\mathcal{K}$ by open sets U_i $(i \in I)$ such that $\rho(U_j)$ is an open subset of a leaf in $\mathcal{F}_D$. This allows us to write $\mathcal{K} = \cup_{\mathcal{L}\in\mathcal{F}_D} V_{\mathcal{L}}$, if we define

$$V_{\mathcal{L}} \equiv \cup\{U_i \mid i \in I,\, \rho(U_i) \subset \mathcal{L}\} \qquad (\mathcal{L} \in \mathcal{F}_D) ,$$

which is an open subset of $\mathcal{K}$. On the other hand, $\mathcal{K} = \cup_{\mathcal{L}\in\mathcal{F}_D} V_{\mathcal{L}}$ is a *disjoint* union, because the leaves in $\mathcal{F}_D$ are (by definition) disjoint. But $\mathcal{K}$ is a leaf and therefore connected. Consequently, for all but one $\mathcal{L} \in \mathcal{F}_D$ the set $V_{\mathcal{L}}$ is empty. That is, $\mathcal{K} = V_{\mathcal{L}}$ for some $\mathcal{L} \in \mathcal{F}_D$. From the definition of $V_{\mathcal{L}}$ it now follows that $\rho(\mathcal{K})$ is a union of open subsets of $\mathcal{L}$, from which the claim above follows.

Now let $\mathcal{L} \in \mathcal{F}_D$ be any leaf. We now prove that the connected components of $\rho^{-1}(\mathcal{L})$ are leaves in $\mathcal{F}_{\rho^*D}$. First, observe that we must have $\rho^{-1}(\mathcal{L}) = \cup_{i\in I} \mathcal{K}_i$ for some collection of leaves $\mathcal{K}_i \in \mathcal{F}_{\rho^*D}$ $(i \in I)$. This follows from the claim proven above, the surjectivity of $\rho : P \to Q$, and the fact that the leaves in $\mathcal{F}_{\rho^*D}$ cover P. Let C be a connected component of $\rho^{-1}(\mathcal{L})$. Then C is a connected integral manifold of D and so is contained in some leaf $\mathcal{K}_i$ $(i \in I)$, by the maximality of leaves (E.5.3). Indeed, by E.5.4, C is an open subset of $\mathcal{K}_i$. Since $\rho^{-1}(\mathcal{L})$ is a disjoint union of its connected components, this shows that for each $i \in I$ the leaf $\mathcal{K}_i$ is a disjoint union of connected components of $\rho^{-1}(\mathcal{L})$. But these components are contained as *open* subsets and $\mathcal{K}_i$ is connected. The only possibility is that $\mathcal{K}_i$ is a single connected component of $\rho^{-1}(\mathcal{L})$.

Suppose now that ρ has connected fibers. It remains to show that $\rho^{-1}(\mathcal{L})$ is a single leaf in $\mathcal{F}_{\rho^*D}$. Write $\rho^{-1}(\mathcal{L}) = \cup_{i\in I} \mathcal{K}_i$ as above. Then $\mathcal{L} = \cup_{i\in I} \rho(\mathcal{K}_i)$. The $\rho(\mathcal{K}_i)$ are open in $\mathcal{L}$ (by above claim) and $\mathcal{L}$ is connected (since $\mathcal{L}$ is a leaf). If we can show that the union $\mathcal{L} = \cup_{i\in I} \rho(\mathcal{K}_i)$ is a disjoint one, then it will follow that the index set I has cardinality one. In that case, $\rho^{-1}(\mathcal{L}) = \mathcal{K}$ for some $\mathcal{K} \in \mathcal{F}_{\rho^*D}$ and we will be done.

To prove disjointness of the union $\mathcal{L} = \cup_{i\in I} \rho(\mathcal{K}_i)$ it clearly suffices to verify the following:

Claim. When ρ has connected fibers, each of these fibers lies entirely within a single leaf in $\mathcal{F}_{\rho^*D}$.

Indeed, fix a point $y \in Q$ and a point $x \in \rho^{-1}(y)$. Since $\ker \mathrm{T}\rho \subset \rho^*D$, it easily follows from the local submersion theorem, and the characterization E.5.2 of leaves, that there exists an open set $U \subset \rho^{-1}(y)$ containing x that lies in a leaf in $\mathcal{F}_{\rho^*D}$. Write $\rho^{-1}(y)$ as a union of such open sets. Then this union is disjoint, because leaves in $\mathcal{F}_{\rho^*D}$ are disjoint. Since $\rho^{-1}(y)$ is connected by hypothesis, it follows that $\rho^{-1}(y)$ is contained in a single leaf in $\mathcal{F}_{\rho^*D}$. This proves the claim and hence concludes the proof of Theorem E.7. □

Bibliography

Abraham, R. and Marsden, J. E. (1978). *Foundations of Mechanics*, 2nd edn, Addison-Wesley Publishing Co., Reading, Massachusetts.

Abraham, R., Marsden, J. E. and Ratiu, T. (1988). *Manifolds, Tensor Analysis, and Applications*, Vol. 75 of *Applied Mathematical Sciences*, 2nd edn, Springer, New York.

Arnold, V. I. (1963). Proof of a theorem of A. N. Kolmogorov on the invariance of quasi-periodic motions under small perturbations of the Hamiltonian, *Russian Math. Surveys* **18 (5)**: 9–36.

Arnold, V. I., Kozlov, V. V. and Neishtadt, A. I. (1988). *Dynamical Systems III*, Encycl. Math. Sci., Springer, New York.

Benettin, G. and Fassò, F. (1996). Fast rotations of the rigid body: a study by Hamiltonian perturbation theory. I., *Nonlinearity* **9**: 137–186.

Benettin, G., Ferrari, G., Galgani, L. and Giorgilli, A. (1982). An extension of the Poincaré-Fermi theorem on the nonexistence of invariant manifolds in nearly integrable Hamiltonian systems, *Nuovo Cimento Soc. Ital. Fis. B* **72**: 137–147.

Blaom, A. D. (2000). Reconstruction phases via Poisson reduction, *Differential Geom. Appl.* **12**(3): 231–252.

Bredon, G. E. (1972). *Introduction to Compact Transformation Groups*, Academic Press, New York.

Bröcker, T. and tom Dieck, T. (1985). *Representations of compact Lie groups*, Vol. 98 of *Graduate Texts in Mathematics*, Springer, New York.

Dazord, P. and Delzant, T. (1987). Le Probleme General des Variables Actions-Angles, *J. Differential Geom.* **26**: 223–251.

Duistermaat, J. J. (1980). On Global Action-Angle Coordinates, *Comm. Pure Appl. Math.* **33**: 687–706.

Fassò, F. (1995). Hamiltonian Perturbation Theory on a Manifold, *Celestial Mech. Dynam. Astronom.* **62**: 43–69.

Giorgilli, A. and Skokos, C. (1997). On the stability of the Trojan asteroids, *Astron. Astrophys.* **317**: 254–261.

Gotay, M. J. (1982). On coisotropic embeddings of presymplectic manifolds, *Proc. Amer. Math. Soc.* **84**: 111–114.

Grauert, H. (1952). On Levi's problem and the imbedding of real-analytic manifolds, *Ann. of Math.* **68**: 460–472.

Guillemin, V., Lerman, E. and Sternberg, S. (1996). *Symplectic Fibrations and Multiplicity Diagrams*, Cambridge University Press.

Guillemin, V. and Sternberg, S. (1984). *Symplectic Techniques in Physics*, Cambridge University Press.

Hilgert, J., Neeb, K.-H. and Plank, W. (1994). Symplectic Convexity Theorems and Coadjoint Orbits, *Compositio Math.* **94**: 129–180.

Kirillov, A. A. (1976). *Elements of the Theory of Representations*, Vol. 220 of *Grundlehren der mathematischen Wissenschaften*, Springer, Berlin.

Kirwan, F. (1984). Convexity properties of the moment mapping III, *Invent. Math.* **77**: 547–552.

Kolmogorov, A. N. (1954). On conservation of conditionally-periodic motions for a small change in Hamilton's function (Russian), *Dokl. Akad. Nauk* **98**(4): 527–530. English translation in *Lecture Notes in Physics* v. 93, G. Casati and J. Ford (Editors), Springer, 1979.

Libermann, P. and Marle, C.-M. (1987). *Symplectic Geometry and Analytical Mechanics*, D. Reidel Publishing Company.

Lichtenberg, A. J. and Lieberman, M. A. (1992). *Regular and Chaotic Dynamics*, Springer, New York.

Lochak, P. (1992). Canonical perturbation theory via simultaneous approximation, *Russian Math. Surveys* **47**(6): 57–133.

Lochak, P. (1993). Hamiltonian perturbation theory: periodic orbits, resonances and intermittency, *Nonlinearity* **6**: 885–904.

Lochak, P. and Meunier, C. (1988). *Multiphase Averaging for Classical Systems*, Vol. 72 of *Applied Mathematical Sciences*, Springer.

Lochak, P. and Neishtadt, A. I. (1992). Estimates of stability time for nearly integrable systems with a quasi-convex Hamiltonian, *Chaos* **2**: 495–499.

Manakov, S. V. (1976). Note on the integration of Euler's equations of the dynamics of an n-dimensional rigid body, *Funct. Anal. Appl.* **1**(4): 93–94.

Marle, C.-M. (1976). Symplectic Manifolds, Dynamical Groups and Hamiltonian Mechanics, *in* M. Cahen and M. Flato (eds), *Differential Geometry and Relativity*, Reidel.

Marle, C.-M. (1982). Sous-variétés de rang constant et sous-variétés symplectiquement régulières d'une variété symplectique, *C. R. Acad. Sci. Paris Sér. I Math.* **295**: 119–122.

Marle, C.-M. (1983a). Normal Forms Generalizing Action-Angle Coordinates for Hamiltonian Actions of Lie Groups, *Letters in Mathematical Physics* **7**: 55–62.

Marle, C.-M. (1983b). Sous-variétés de rang constant d'une variété symplectique, *Astérisque* **107–108**: 69–86.

Marsden, J. E. (1981). *Lectures on Geometric Methods in Mathematical Physics*, Vol. 37 of *CBMS*, SIAM, Philadelphia.

Marsden, J. E. (1992). *Lectures on Mechanics*, Vol. 174 of *London Mathematical Society Lecture Notes Series*, Cambridge University Press.

Marsden, J. E., Montgomery, R. and Ratiu, T. (1990). Reduction, symmetry, and phases in mechanics, *Mem. Amer. Math. Soc.* **88**(436): 1–110.

Marsden, J. E. and Ratiu, T. S. (1994). *Introduction to Mechanics and Symmetry*, Vol. 17 of *Texts in Applied Mathematics*, Springer.

Marsden, J. E. and Weinstein, A. (1974). Reduction of symplectic manifolds with symmetry, *Rep. Math. Phys.* **5**: 121–130.

Melnikov, V. K. (1963). On the stability of the center for time periodic perturbations, *Trans. Moscow Math. Soc.* **12**: 1–57.

Meyer, K. R. (1973). Symmetries and Integrals in Mechanics, *in* M. Peixoto (ed.), *Dynamical Systems*, Academic Press, New York, pp. 259–273.

Mishchenko, A. S. and Fomenko, A. T. (1978). Generalized Liouville method of integration of Hamiltonian systemssymplectic data, *Funct. Anal. Appl.* **12**: 113–121.

Moser, J. (1962). On invariant curves of area preserving mappings of an annulus, *Nachr. Akad. Wiss. Göttingen Math.-Phys. K1. II* **a (1)**: 1–20.

Moser, J. (1965). On the volume elements on a manifold, *Trans. Amer. Math. Soc.* **120**: 286–294.

Moser, J. (1970). Regularization of Kepler's problem and the averaging method on a manifold, *Comm. Pure Appl. Math.* **23**: 609–636.

Neishtadt, A. I. (1984). On the separation of motions in systems with rapidly rotating phase, *J. Appl. Math. Mech.* **48**: 133–139.

Nekhoroshev, N. N. (1972). Action-Angle Variables and their Generalizations, *Trans. Moscow Math. Soc.* **26**: 180–198.

Nekhoroshev, N. N. (1977). An Exponential Estimate of the Time of Stability Of Nearly-Integrable Hamiltonian Systems, *Russian Math. Surveys* **32 (6)**: 5–66.

Ortega, J.-P. (1998). *Symmetry, Reduction, and Stability in Hamiltonian Systems*, PhD thesis, University of California, Santa Cruz. Available at `http://dmawww.epfl.ch/~ortega`.

Schmidt, W. M. (1980). *Diophantine Approximation*, Vol. 785 of *Lecture Notes in Mathematics*, Springer, Berlin.

Sharpe, R. W. (1997). *Differential Geometry: Cartan's generalization of Klein's Erlangen program*, Vol. 166 of *Graduate Texts in Mathematics*, Springer.

Stefan, P. (1973). Accessible sets, orbits, and foliations with singularities, *Proc. London Math. Soc.* **29**: 699–713.

Sussmann, H. J. (1973). Orbits of families of vector fields and integrability of distributions, *Trans. Amer. Math. Soc.* **180**: 171–188.

Tantalo, P. (1994). *Geometric Phases for the Free Rigid Body with Variable Inertia Tensor*, PhD thesis, University of California, Santa Cruz.

Weinstein, A. (1971). Symplectic manifolds and their Lagrangian submanifolds, *Adv. Math.* **6**: 329–346.

Weinstein, A. (1981). Neighborhood classification of isotropic embeddings, *J. Differential Geom.* **16**: 125–128.

Weinstein, A. (1983). The local structure of Poisson manifolds, *J. Differential Geom.* **18**: 523–557.

Whitney, H. and Bruhat, F. (1959). Quelques propriétés fondamentales des ensembles analytiques-réels, *Comment. Math. Helv.* **33**: 132–160.

Editorial Information

To be published in the *Memoirs*, a paper must be correct, new, nontrivial, and significant. Further, it must be well written and of interest to a substantial number of mathematicians. Piecemeal results, such as an inconclusive step toward an unproved major theorem or a minor variation on a known result, are in general not acceptable for publication. Papers appearing in *Memoirs* are generally longer than those appearing in *Transactions*, which shares the same editorial committee.

As of May 31, 2001, the backlog for this journal was approximately 7 volumes. This estimate is the result of dividing the number of manuscripts for this journal in the Providence office that have not yet gone to the printer on the above date by the average number of monographs per volume over the previous twelve months, reduced by the number of volumes published in four months (the time necessary for preparing a volume for the printer). (There are 6 volumes per year, each containing at least 4 numbers.)

A Consent to Publish and Copyright Agreement is required before a paper will be published in the *Memoirs*. After a paper is accepted for publication, the Providence office will send a Consent to Publish and Copyright Agreement to all authors of the paper. By submitting a paper to the *Memoirs*, authors certify that the results have not been submitted to nor are they under consideration for publication by another journal, conference proceedings, or similar publication.

Information for Authors

Memoirs are printed from camera copy fully prepared by the author. This means that the finished book will look exactly like the copy submitted.

The paper must contain a *descriptive title* and an *abstract* that summarizes the article in language suitable for workers in the general field (algebra, analysis, etc.). The *descriptive title* should be short, but informative; useless or vague phrases such as "some remarks about" or "concerning" should be avoided. The *abstract* should be at least one complete sentence, and at most 300 words. Included with the footnotes to the paper should be the 2000 *Mathematics Subject Classification* representing the primary and secondary subjects of the article. The classifications are accessible from `www.ams.org/msc/`. The list of classifications is also available in print starting with the 1999 annual index of *Mathematical Reviews*. The Mathematics Subject Classification footnote may be followed by a list of *key words and phrases* describing the subject matter of the article and taken from it. Journal abbreviations used in bibliographies are listed in the latest *Mathematical Reviews* annual index. The series abbreviations are also accessible from `www.ams.org/publications/`. To help in preparing and verifying references, the AMS offers MR Lookup, a Reference Tool for Linking, at `www.ams.org/mrlookup/`. When the manuscript is submitted, authors should supply the editor with electronic addresses if available. These will be printed after the postal address at the end of the article.

Electronically prepared manuscripts. The AMS encourages electronically prepared manuscripts, with a strong preference for AMS-LaTeX. To this end, the Society has prepared AMS-LaTeX author packages for each AMS publication. Author packages include instructions for preparing electronic manuscripts, the *AMS Author Handbook*, samples, and a style file that generates the particular design specifications of that publication series. Though AMS-LaTeX is the highly preferred format of TeX, author packages are also available in AMS-TeX.

Authors may retrieve an author package from e-MATH starting from `www.ams.org/tex/` or via FTP to `ftp.ams.org` (login as `anonymous`, enter username as password, and type `cd pub/author-info`). The *AMS Author Handbook* and the *Instruction Manual* are available in PDF format following the author packages link from `www.ams.org/tex/`. The author package can be obtained free of charge by sending email to `pub@ams.org` (Internet) or from the Publication Division, American Mathematical Society, P.O. Box 6248, Providence, RI 02940-6248. When requesting an author package, please specify AMS-LaTeX or AMS-TeX, Macintosh or IBM (3.5) format, and the publication in which your paper will appear. Please be sure to include your complete mailing address.

Sending electronic files. After acceptance, the source file(s) should be sent to the Providence office (this includes any TeX source file, any graphics files, and the DVI or PostScript file).

Before sending the source file, be sure you have proofread your paper carefully. The files you send must be the EXACT files used to generate the proof copy that was accepted for publication. For all publications, authors are required to send a printed copy of their paper, which exactly matches the copy approved for publication, along with any graphics that will appear in the paper.

TeX files may be submitted by email, FTP, or on diskette. The DVI file(s) and PostScript files should be submitted only by FTP or on diskette unless they are encoded properly to submit through email. (DVI files are binary and PostScript files tend to be very large.)

Electronically prepared manuscripts can be sent via email to `pub-submit@ams.org` (Internet). The subject line of the message should include the publication code to identify it as a Memoir. TeX source files, DVI files, and PostScript files can be transferred over the Internet by FTP to the Internet node `e-math.ams.org` (130.44.1.100).

Electronic graphics. Comprehensive instructions on preparing graphics are available at `www.ams.org/jourhtml/graphics.html`. A few of the major requirements are given here.

Submit files for graphics as EPS (Encapsulated PostScript) files. This includes graphics originated via a graphics application as well as scanned photographs or other computer-generated images. If this is not possible, TIFF files are acceptable as long as they can be opened in Adobe Photoshop or Illustrator. No matter what method was used to produce the graphic, it is necessary to provide a paper copy to the AMS.

Authors using graphics packages for the creation of electronic art should also avoid the use of any lines thinner than 0.5 points in width. Many graphics packages allow the user to specify a "hairline" for a very thin line. Hairlines often look acceptable when proofed on a typical laser printer. However, when produced on a high-resolution laser imagesetter, hairlines become nearly invisible and will be lost entirely in the final printing process.

Screens should be set to values between 15% and 85%. Screens which fall outside of this range are too light or too dark to print correctly. Variations of screens within a graphic should be no less than 10%.

Inquiries. Any inquiries concerning a paper that has been accepted for publication should be sent directly to the Electronic Prepress Department, American Mathematical Society, P. O. Box 6248, Providence, RI 02940-6248.

Editors

This journal is designed particularly for long research papers, normally at least 80 pages in length, and groups of cognate papers in pure and applied mathematics. Papers intended for publication in the *Memoirs* should be addressed to one of the following editors. In principle the Memoirs welcomes electronic submissions, and some of the editors, those whose names appear below with an asterisk (*), have indicated that they prefer them. However, editors reserve the right to request hard copies after papers have been submitted electronically. Authors are advised to make preliminary email inquiries to editors about whether they are likely to be able to handle submissions in a particular electronic form.

Algebra to CHARLES CURTIS, Department of Mathematics, University of Oregon, Eugene, OR 97403-1222 email: `cwc@darkwing.uoregon.edu`

Algebraic geometry and commutative algebra to LAWRENCE EIN, Department of Mathematics, University of Illinois, 851 S. Morgan (M/C 249), Chicago, IL 60607-7045; email: `ein@uic.edu`

Algebraic topology and cohomology of groups to STEWART PRIDDY, Department of Mathematics, Northwestern University, 2033 Sheridan Road, Evanston, IL 60208-2730; email: `priddy@math.nwu.edu`

Combinatorics and Lie theory to SERGEY FOMIN, Department of Mathematics, University of Michigan, Ann Arbor, Michigan 48109-1109; email: `fomin@math.lsa.umich.edu`

Complex analysis and complex geometry to DUONG H. PHONG, Department of Mathematics, Columbia University, 2990 Broadway, New York, NY 10027-0029; email: `phong@math.columbia.edu`

***Differential geometry and global analysis** to LISA C. JEFFREY, Department of Mathematics, University of Toronto, 100 St. George St., Toronto, ON Canada M5S 3G3; email: `jeffrey@math.toronto.edu`

***Dynamical systems and ergodic theory** to ROBERT F. WILLIAMS, Department of Mathematics, University of Texas, Austin, Texas 78712-1082; email: `bob@math.utexas.edu`

Functional analysis and operator algebras to BRUCE E. BLACKADAR, Department of Mathematics, University of Nevada, Reno, NV 89557; email: `bruceb@math.unr.edu`

Geometric topology, knot theory and hyperbolic geometry to ABIGAIL A. THOMPSON, Department of Mathematics, University of California, Davis, Davis, CA 95616-5224; email: `thompson@math.ucdavis.edu`

Harmonic analysis, representation theory, and Lie theory to ROBERT J. STANTON, Department of Mathematics, The Ohio State University, 231 West 18th Avenue, Columbus, OH 43210-1174; email: `stanton@math.ohio-state.edu`

***Logic** to THEODORE SLAMAN, Department of Mathematics, University of California, Berkeley, CA 94720-3840; email: `slaman@math.berkeley.edu`

Number theory to MICHAEL J. LARSEN, Department of Mathematics, Indiana University, Bloomington, IN 47405; email: `larsen@math.indiana.edu`

***Ordinary differential equations, partial differential equations, and applied mathematics** to PETER W. BATES, Department of Mathematics, Brigham Young University, 292 TMCB, Provo, UT 84602-1001; email: `peter@math.byu.edu`

***Partial differential equations and applied mathematics** to BARBARA LEE KEYFITZ, Department of Mathematics, University of Houston, 4800 Calhoun Road, Houston, TX 77204-3476; email: `keyfitz@uh.edu`

***Probability and statistics** to KRZYSZTOF BURDZY, Department of Mathematics, University of Washington, Box 354350, Seattle, Washington 98195-4350; email: `burdzy@math.washington.edu`

***Real and harmonic analysis and geometric partial differential equations** to WILLIAM BECKNER, Department of Mathematics, University of Texas, Austin, TX 78712-1082; email: `beckner@math.utexas.edu`

All other communications to the editors should be addressed to the Managing Editor, WILLIAM BECKNER, Department of Mathematics, University of Texas, Austin, TX 78712-1082; email: `beckner@math.utexas.edu`.

Selected Titles in This Series

(Continued from the front of this publication)

For a complete list of titles in this series, visit the AMS Bookstore at **www.ams.org/bookstore/**.